Current Topics in Microbiology and Immunology

Volume 329

Series Editors

Richard W. Compans
Emory University School of Medicine, Department of Microbiology and
Immunology, 3001 Rollins Research Center, Atlanta, GA 30322, USA

Max D. Cooper
Department of Pathology and Laboratory Medicine, Georgia Research Alliance,
Emory University, 1462 Clifton Road, Atlanta, GA 30322, USA

Tasuku Honjo
Department of Medical Chemistry, Kyoto University, Faculty of Medicine,
Yoshida, Sakyo-ku, Kyoto 606-8501, Japan

Hilary Koprowski
Thomas Jefferson University, Department of Cancer Biology, Biotechnology
Foundation Laboratories, 1020 Locust Street, Suite M85 JAH, Philadelphia,
PA 19107-6799, USA

Fritz Melchers
Biozentrum, Department of Cell Biology, University of Basel, Klingelbergstr.
50–70, 4056 Basel Switzerland

Michael B.A. Oldstone
The Scripps Research Institute, Department of Immunology and Microbial
Science, 10550 N. Torrey Pines, La Jolla, CA 92037, USA

Sjur Olsnes
Department of Biochemistry, Institute for Cancer Research, The Norwegian
Radium Hospital, Montebello 0310 Oslo, Norway

Peter K. Vogt
The Scripps Research Institute, Dept. of Molecular & Exp. Medicine, Division of
Oncovirology, 10550 N. Torrey Pines. BCC-239, La Jolla, CA 92037, USA

Diane E. Griffin • Michael B.A. Oldstone
Editors

Measles

History and Basic Biology

 Springer

Editors:
Diane E. Griffin
Johns Hopkins University
School of Hygiene and Public Health
Department of Molecular Microbiology
615 N. Wolfe Street
Baltimore, MD 21205
USA
dgriffin@mail.jhmi.edu
dgriffin@jhsph.edu

Michael B.A. Oldstone
Scripps Research Institute
Department of Immunology and
 Microbial Science
10550 N. Torrey Pines Road
La Jolla, CA 92037
USA
mbaobo@scripps.edu

ISBN 978-3-540-70522-2   e-ISBN 978-3-540-70523-9
DOI 10.1007/978-3-540-70523-9

Current Topics in Microbiology and Immunology ISSN 0070-217X

Library of Congress Catalog Number: 2008931704

© 2009 Springer-Verlag Berlin Heidelberg

This work is subject to copyright. All rights reserved, whether the whole or part of the material is concerned, specifically the rights of translation, reprinting, reuse of illustrations, recitation, broadcasting, reproduction on microfilm or in any other way, and storage in data banks. Duplication of this publication or parts thereof is permitted only under the provisions of the German Copyright Law of September, 9, 1965, in its current version, and permission for use must always be obtained from Springer-Verlag. Violations are liable for prosecution under the German Copyright Law.

The use of general descriptive names, registered names, trademarks, etc. in this publication does not imply, even in the absence of a specific statement, that such names are exempt from the relevant protective laws and regulations and therefore free for general use.

Product liability: The publisher cannot guarantee the accuracy of any information about dosage and application contained in this book. In every individual case the user must check such information by consulting the relevant literature.

Cover design: WMXDesign GmbH, Heidelberg, Germany

Printed on acid-free paper

9 8 7 6 5 4 3 2 1

springer.com

Contents

Introduction 1

1 John F. Enders and Measles Virus Vaccine—a Reminiscence 3
   S.L. Katz

2 Measles Virus Receptors 13
   Y. Yanagi, M. Takeda, S. Ohno, and T. Hashiguchi

3 Measles Virus and CD46 31
   C. Kemper and J.P. Atkinson

4 Measles Virus Glycoprotein Complex Assembly,
   Receptor Attachment, and Cell Entry 59
   C.K. Navaratnarajah, V.H.J. Leonard, and R. Cattaneo

5 The Measles Virus Replication Cycle 77
   B.K. Rima and W.P. Duprex

6 Nucleocapsid Structure and Function 103
   S. Longhi

7 Reverse Genetics of Measles Virus and Resulting
   Multivalent Recombinant Vaccines: Applications
   of Recombinant Measles Viruses 129
   M.A. Billeter, H.Y. Naim, and S.A. Udem

8 Measles Virus Interaction with Host Cells
   and Impact on Innate Immunity 163
   D. Gerlier and H. Valentin

Index 193

Contributors

J.P. Atkinson
Division of Rheumatology, 660 S Euclid, Box 8045, St. Louis, MO 63110, USA, jatkinso@im.wustl.edu

M. Billeter
University of Zurich, Winterthurerstrasse 190, 8057 Zurich, Switzerland, billeter@access.unizh.ch

R. Cattaneo
Mayo Clinic College of Medicine, Dept. of Molecular Medicine, Virology and Gene Therapy Graduate Track, 200 1st St SW, Guggenheim 1838, Rochester MN 55905, USA, Cattaneo.Roberto@mayo.edu

W.P. Duprex
Division of Infection and Immunity, Centre for Cancer Research and Cell Biology, School of Biomedical Sciences, Queen's University Belfast, Belfast BT9 7BL, Northern Ireland, UK

D. Gerlier
Interactions virus cellule-hôte; CNRS; Université de Lyon 1; FRE3011, IFR 62 Laennec, 69372 Lyon Cedex 08, France, denis.gerlier@univ-lyon1.fr

T. Hashiguchi
Department of Virology, Faculty of Medicine, Kyushu University, Fukuoka 812-8582, Japan

S.L. Katz
Box 2925, Duke University Medical Center, Durham, NC 27710, USA, katz0004@mc.duke.edu

C. Kemper
Division of Rheumatology, 660 S Euclid, Box 8045, St. Louis, MO 63110, USA

V. Leonard
Mayo Clinic College of Medicine, Dept. of Molecular Medicine, Virology and Gene Therapy Graduate Track, 200 1st St SW, Guggenheim 1838, Rochester MN 55905, USA

S. Longhi
Architecture et Fonction des Macromolécules Biologiques, UMR 6098 CNRS et Universités Aix-Marseille I et II, 163, avenue de Luminy, Case 932, 13288 Marseille Cedex 09, France, Sonia.Longhi@afmb.univ-mrs.fr

H.Y. Naim
Berna Biotech, Rehhagstrasse 79, 3018 Berne, Switzerland

C. Navaratnarajah
Mayo Clinic College of Medicine, Dept. of Molecular Medicine, Virology and Gene Therapy Graduate Track, 200 1st St SW, Guggenheim 1838, Rochester, MN 55905, USA

S. Ohno
Department of Virology, Faculty of Medicine, Kyushu University, Fukuoka 812-8582, Japan

B.K. Rima
Centre for Infection and Immunity, School of Medicine, Dentistry and Biomedical Sciences, Queen's University Belfast, Belfast BT9 7BL, Northern Ireland, UK, b.rima@qub.ac.uk

M. Takeda
Department of Virology, Faculty of Medicine, Kyushu University, Fukuoka 812-8582, Japan

S. Udem
Consultant for Internatl. AIDS Vaccine Initiative (IAVI), 155W 70th Street, Apt. 6F/G, New York, NY 10023, USA

H. Valentin
Interaction virus-système immunitaire; INSERM; U851, IFR128 BioSciences Lyon-Gerland, Université de Lyon 1, UCBL1 (Lyon, 69003); Hospices Civils de Lyon (Lyon, 69002); 21 Avenue Tony Garnier, 69365 Lyon Cedex 07, France

Y. Yanagi
Department of Virology, Faculty of Medicine, Kyushu University, Fukuoka 812-8582, Japan, yyanagi@virology.med.kyushu-u.ac.jp

Introduction

Measles virus, one of the most contagious of all human viruses, has been largely contained by the development and use of a vaccine that was introduced 50 years ago. These two volumes were timed to honor the introduction of the vaccine and to record the enormous advancements made in understanding the molecular and cell biology, pathogenesis, and control of this infectious disease. Where vaccine has been effectively delivered, endemic measles virus transmission has been eliminated. However, difficulties in vaccine delivery, lack of health care support and objection to vaccination in some communities continue to result in nearly 40 million cases and over 300,000 deaths per year from measles.

By itself measles virus infection has and still provides some of the most interesting phenomena in biology. Following infection of dendritic cells, measles virus causes a profound suppression of the host's immune response that lasts a number of months after apparent recovery from infection. Indeed, measles virus was the first virus to be associated with immunosuppression with many of the manifestations to be observed one hundred years later with HIV infection. Measles is also associated with development of both post-infectious encephalomyelitis, an autoimmune demyelinating disease, and subacute sclerosing panencephalitis, a slowly progressive neurodegenerative disorder. How measles virus infects cells, spreads to various tissues and causes disease, as well as the role of the immune response, generation of new vaccines, and use as a vector for gene delivery are topics covered in these two volumes. A unique highlight for readers of this series and those interested in the history of a major and profound biomedical research accomplishment is the chapter written by one of the participants who worked on the initial discovery and use of the vaccine who records the events that occurred at that time.

Baltimore, MD
La Jolla, CA

Diane E. Griffin
Michael B.A. Oldstone

Chapter 1
John F. Enders and Measles Virus Vaccine—a Reminiscence

S.L. Katz

Contents

References 10

Abstract Following their initial isolation in cell culture of the virus in 1954, a succession of investigators under the mentorship of John F. Enders conducted the research, development, and initial clinical studies responsible for the licensure in 1963 of a successful live attenuated measles virus vaccine. Propagation of the virus successively in human kidney cells, human amnion cells, embryonated hens' eggs, and finally chick embryo cell cultures had selected virus that when inoculated into susceptible monkeys proved immunogenic without viremia or overt disease, in contrast to the early kidney cell-passaged material, which in similar monkeys produced viremia with illness mimicking human measles. Careful clinical studies in children by the Enders group and then by collaborating investigators in many sites established its safety, immunogenicity, and efficacy. This Edmonston strain measles virus became the progenitor of vaccines prepared, studied, and utilized throughout the United States and many other countries. With appreciation of measles morbidity and mortality, most marked among infants and children in the resource-limited lands, the vaccine was incorporated into the World Health Organization's (WHO) Expanded Programme of Immunization (EPI) in 1974 along with BCG, OPV, and DTP. Successful efforts to further reduce measles' burden were launched in 2001 and are continuing as the Measles Initiative (Partnership) under the leadership of the American Red Cross, International Red Cross, and Red Crescent societies, Centers for Disease Control (CDC), United Nations Children's Fund (UNICEF), WHO, and the United Nations Foundation.

S.L. Katz
Duke University Medical Center, Box 2925, Durham, NC 27710, USA, e-mail: katz0004@mc.duke.edu

When in 1954 John F. Enders and his two younger colleagues, Frederick Robbins and Thomas Weller, received the Nobel Prize in Physiology or Medicine for the cultivation in cell cultures of polio viruses, he had already returned to his initial interest in isolating and propagating the virus responsible for measles. Enders was a unique investigator whose career had followed a less than conventional path. The scion of a wealthy Connecticut Yankee family, he had attended an elite boys preparatory school, St. Paul's in Concord, New Hampshire, and then Yale University. Additionally, he had spent an interim year as a US Navy World War I flight instructor. Upon university graduation, he was provided a position in the family's banking enterprise, responsible for selling real estate. Recognizing his lack of interest and commitment to such a pursuit, he enrolled in the Graduate School at Harvard University studying ancient Celtic philology. A fortunate turn of events provided him a roommate in their rented Brookline apartment, Hugh Ward, a budding Australian microbiologist. Ward was apprenticed to Hans Zinsser, the eminent microbiologist, in whose laboratory he studied. Enders visited the laboratory where he became fascinated by the projects of his roommate and the views he gained through Ward's microscope. Abandoning Celtic philology, he joined Zinsser's group as a graduate student in microbiology. Here began the career for which he is widely remembered and appreciated. His early work focused on the role of complement and antibody in the response to pneumococcal infections. After, and perhaps because of, the death of his first wife from influenza virus infection, he directed his investigative efforts to viruses. Early work involved feline panleukopenia, a fatal disease of cats, mumps virus, and then polio.

Enders was a very special individual, not fitting the mold of the aggressive goal-oriented researcher, but more a contemplative, broadly interested investigator who pursued medical science for enrichment of the field and personal gratification, but not for audience acclaim. He was an ideal mentor for many young aspirants, most of whom later succeeded in developing subsequent careers as distinguished scientists. Because he believed that daily and leisurely contact with one's disciples was critical to their advancement and the productivity of his laboratory, he never accepted more than four or five fellows at any time, a marked contrast to many of his contemporaries. In addition to those from the United States, he enjoyed opening his laboratory to bright young fellows from abroad (Japan, Iran, the Netherlands, Sweden, England, Yugoslavia, Belgium, Germany, South Africa, Turkey). On the daily rounds of the laboratory benches, his question "What's new?" provided an effective stimulus to each fellow to have developed something that would then catch his interest, initiating then a 30- or 60-minute conversation in which the significance of the findings and the ways in which one might pursue further studies were discussed. One worked *with* John Enders not *for* him (Table 1).

In 1954, one pediatric fellow who spent a year in the laboratory was dispatched to a suburban school where an outbreak of measles was reported to be underway. Thomas Peebles obtained throat swabs and blood specimens from the affected youngsters and brought them back to the laboratory. In conventional Enders' fashion, never to waste material and always to utilize available opportunities, cells from human kidneys had been successfully cultured in vitro. These originated from a

Table 1 Enders laboratory participants in the research and development of measles virus vaccine

Thomas Peebles	Samuel Katz
Kevin McCarthy	Ann Holloway
Anna Mitus	Donald Medearis
Milan Milovanovic	Elizabeth Grogan

neurosurgical procedure then in fashion in which children with hydrocephalus had a unilateral nephrectomy with a connection then established between the cerebrospinal fluid in the subarachnoid space and the ureter of the sacrificed kidney. These kidneys came to the Enders' lab, where they were minced, trypsinized, and put into cell culture with nutrient media. It was in these cells that measles virus was first successfully cultivated and passaged a number of times (Enders and Peebles 1954). Because the name of the young student from whom the original virus had been isolated was David Edmonston, this strain of virus has subsequently always been identified as Edmonston virus. Virus harvested from early passage of these cultures was inoculated into measles-susceptible monkeys who then developed fever, rash, viremia, and eventually measles-specific antibodies, both complement-fixing and virus-neutralizing (Peebles et al. 1957). As ventriculoureteral shunts fell out of fashion, with the development of improved technology for relief of hydrocephalus, new sources of human cells were sought. At a neighboring hospital, an obstetrical institution, 15–20 women were delivered each day of newborns and their placentas cast aside. Enders, in his customarily frugal but innovative fashion, suggested we strip the amniotic membrane from these discarded placentas and attempt to prepare cultures of human amnion cells. One of the fellows went to the obstetrical hospital to claim a placenta, brought it back to the laboratory, where it was mounted so that the amniotic membrane could be sterilely removed. The membrane was then trypsinized, and the resultant cells were dispersed, harvested, and placed in test tubes and flasks where they were successfully grown. Measles virus after 24 passages in human kidney cells replicated effectively in these human amnion cell cultures and once again produced an identifiable cytopathic effect (Milovanovic et al. 1957). With typical Enders' imaginative approach, he then suggested that if the virus grew readily in human amnion cells, perhaps it would also replicate in a nonhuman but similar environment. Therefore after 28 human amnion cell passages, we moved to embryonated hen's eggs and inoculated virus intra-amniotically. The eggs were obtained from a supposedly pathogen-free flock in New Hampshire. Although there was no visible resultant pathology, fluids harvested from these infected eggs displayed cytopathology when inoculated back into human amnion cells, and titers indicated the virus had not merely persisted but had multiplied (Milovanovic et al. 1957). After six passages in the fertile hen's eggs, we prepared cell cultures from trypsinized chick embryo tissue and inoculated virus into those tubes. Although no effect was seen for the initial passages, after five, there was visible cytopathology which coincided with demonstrable replication of the virus in these cultures (Katz et al. 1958). It was 13th passaged chick cell material that was inoculated into measles-susceptible monkeys and the results compared with the original early human kidney-cell-propagated virus. In contrast, the chick cell virus produced no rash, no detectable viremia but nonetheless complement-fixing and virus-neutralizing antibodies (Enders et al. 1960). In addition to the aforementioned studies, chick cell virus was also inoculated directly into the cistern and the cerebral hemispheres of susceptible monkeys. No behavioral changes were noted after this procedure, but the animals were sacrificed and neuropathological studies of the infected cerebral tissue were conducted by veterinary pathologists at the neighboring animal hospital. No histological changes could be identified. In contrast, monkeys similarly injected intracranially with early passaged kidney cell virus developed lesions with local mononuclear cell infiltrates, perivascular cuffing, and demyelination. In another series of experiments, monkeys that had been immunized with the chick cell virus were then challenged with the virulent human kidney cell virus and proved completely resistant to infection. After these successful studies in monkeys, the question was how next to proceed to evaluation in humans.

Initially, we prepared lots of serum-free vaccine virus carefully scrutinized and tested for any contaminating agents and for sterility, to inoculate one another. Although this was not a test of efficacy, it was a determinant of possible toxicity and safety. With the successful completion of these preliminary studies, we then considered how best to proceed to study the vaccine in susceptible children. At a nearby state Institution for physically and intellectually challenged youngsters, outbreaks of measles occurred every 2 or 3 years, resulting in serious morbidity and a number of deaths. Following discussions with the institutional director, we were able to meet with the parents of several dozen children who had not yet suffered measles. After explaining to them the background of our potential vaccine and our plans for a clinical trial, most of them agreed to have their children participate. Using the same materials with which we had inoculated one another in the laboratory, we proceeded to inject subcutaneously a dozen susceptible children with the vaccine and several with sterile tissue culture fluid as placebo. We examined them daily, obtained nasopharyngeal cultures and venous blood samples on alternate days and followed them carefully over the next 3 weeks. Five to 8 days after inoculation, many of them developed fevers that persisted for several days and were then followed by an evanescent rash. Throughout this time, they nevertheless remained well and went about their normal activities. No virus was recovered from the throat cultures or blood, but within 2 weeks all had detectable measles virus-neutralizing and complement-fixing antibodies in their sera (Katz and Enders 1959; Katz et al. 1960a). The nursing personnel and others responsible for these children attested to the absence of any apparent disability during this time. Buoyed by these initially successful studies, we enlisted colleagues in Denver, New Haven, Cleveland, New York, and Boston to conduct similar studies among home-dwelling children under their care. The successful completion of these studies resulted in the *New England Journal of Medicine* reports in 1960 describing the background and development of the vaccine virus and the clinical observations of the vaccinated children (Katz et al. 1960b).

Throughout the years of this laboratory and clinical research (1954–1963), the Enders laboratory made available to any and all legitimate investigators who were interested in pursuing related studies varied materials for their use. These included virus, cell cultures, and sera. The Enders philosophy was that the more people working on a problem the sooner solutions would be found. There was never any intent to patent the virus or to seek monetary return. As a result, within a short period of time, many university groups pursued measles vaccine investigations and seven different pharmaceutical firms in the United States and several abroad were producing their versions of the Edmonston measles virus vaccine (Table 2; Fig. 1). To attenuate further the clinical results of the initial vaccination (the aforementioned fever and exanthem), protocols were initiated in which the injection of vaccine was accompanied by a simultaneous tiny dose of human immunoglobulin (0.02 mg/kg body weight), which reduced these manifestations to approximately

Table 2 US Firms that produced measles virus vaccines

Pfizer	Lilly
Parke-Davis	Lederle
Philips-Roxane	Pitman Moore-Dow
Merck (Sharpe and Dohme)[a]	

[a] The sole remaining US producer

Fig. 1 First International Conference on Measles Immunization. 8 November 1961 at the National Institutes of Health, Bethesda Maryland. *Left to right*: Samuel Katz, Ann Holloway, Kevin McCarthy, Anna Mitus, Milan Milovanovic, John Enders, Gisele Ruckle, Frederick Robbins, Ikuyu Nagata

10%–15%t of susceptible recipients. A number of investigators (initially Anton Schwarz in 1965 at American Home Products-Pittman Moore Dow and later Maurice Hilleman in 1968 at Merck) further attenuated the Edmonston virus by an increased number of passages in chick embryo fibroblasts at reduced temperature (32°C in contrast to the usual 35°–36°C). Additionally, several firms prepared formalin-inactivated, alum-precipitated measles vaccine from the Edmonston strain and studied its use in a three-dose schedule (Rauh and Schmidt 1965). The Enders group remained committed to live vaccine, convinced of its advantages over the inactivated preparation (Enders et al. 1962). Both the live attenuated and this inactivated vaccine were licensed in the United States on 21 March 1963. Over the ensuing several years, it was discovered that the killed vaccine did not produce enduring immunity and that when recipients were exposed to wild measles, many developed a severe atypical measles infection characterized by high fever, unusual rash beginning most prominently on the extremities, pneumonia with residual pulmonary nodules, and some central nervous system obtundation (Fulginiti et al. 1967; Annunziato et al. 1982). This inactivated vaccine was therefore withdrawn from use in 1967.

Fortunately it was not until 1969, 6 years after the licensure of measles virus vaccines, that the responsibility of wild measles virus for subacute sclerosing panencephalitis (SSPE) was discovered (Horta-Barbosa et al. 1969; Payne et al. 1969). By then, millions of American children had received live-attenuated measles virus vaccines with no resultant central nervous system complications resembling SSPE, and annual measles cases had been reduced by more than 90%. If the association of measles with SSPE had been appreciated prior to 1963, it is questionable whether licensure of a live-virus vaccine would have been so readily approved. Reassuringly, not only has SSPE become an extreme rarity in the United States and other countries with widespread childhood coverage by measles vaccine, but Bellini and colleagues at CDC have demonstrated that all the few cases identified in recent years are attributable genotypically to wild-type virus distinct from the vaccine strain (Bellini et al. 2005).

Early in development of the vaccine, after several presentations at national and international meetings, we began to receive a number of communications from Dr. David Morley, a British pediatrician who was developing child health programs in Nigeria, where he informed us that mortality from measles frequently approached 10%–20%. Of 555 children at his clinic 125 died of measles! He urged us to come to Nigeria and study the vaccine there. Judiciously, however, John Enders cautioned us to wait until the vaccine had proven its safety and efficacy in US youngsters before embarking on such a mission. His concern was that premature studies would be regarded as taking advantage of human guinea pigs rather than as a humane medical mission. Responding eventually to Morley's entreaties, Katz went in 1960 with Edmonston vaccine provided by Merck, which was then involved in its initial commercial production. The clinical trial was conducted in Imesi-ile, a tiny village outside Ilesha, a larger market town north of Ibadan. When informed of the project, local mothers keenly aware of measles' morbidity and mortality, eagerly brought their infants and children to participate. Many of these youngsters had malaria, protein malnutrition, and intestinal nematode infestations. Despite these severe compromises, the initial 26 recipients responded favorably to the vaccine, had no adverse events, and developed antibodies at the expected time (Katz et al. 1962). A secondary benefit of this experience was our personal awakening to an awareness of the serious morbidity and mortality of measles among infants and children in the resource-limited nations. Our previous perspective had been a rather parochial one, of measles in the United States where nearly every child by age 7 had acquired the infection. Complications including otitis media, pneumonia, and gastroenteritis were common, requiring hospitalization in as many as 20%, but mortality was unusual, approximately one in 500 cases. Progress in the Americas had been remarkably successful, with transmission in the United States halted in 1993 (Katz and Hinman 2004) and in the entire Western hemisphere by 2002 (de Quadros et al. 2004). The few cases identified since then have been attributable to importations from countries where measles remains endemic. Although initial success in control was mainly the result of a single dose schedule, it became apparent that the 5%–10% of recipients who failed to seroconvert after this administration soon constituted a significant cluster of susceptibles in whom such a highly transmissible virus could ignite an outbreak. Therefore, beginning in the early 1990s, a two-dose schedule became the routine and has been continued worldwide in those nations where measles vaccination is practiced.

The experience in Nigeria stimulated our endeavors to place measles vaccine on the global scene, resulting eventually in its inclusion in the Expanded Program on Immunization (EPI) of the World Health Organization (WHO). However, there were still millions of deaths each year and no international effort was initiated, whereas the global focus was on polio eradication (Katz 2005). However, by the year 2000, the American Red Cross and International Red Cross and Red Crescent Societies, joined by the Centers for Disease Control (CDC), the United Nations Children's Fund (UNICEF), the United Nations Foundation, and the World Health Organization (WHO), formed the Measles Partnership (Measles Initiative) with its goal of reducing measles mortality from 873,000 annually (WHO figures for 1999) to half in the next 5 years. Remarkably, in the initial 5 years they exceeded their goal with vaccination of 297 million infants and children (ages 9 months to 5 years) and a resultant 68% overall decrease in measles mortalities (Wolfson et al. 2007). Most of this was in sub-Saharan Africa, where only 126,000 deaths were recorded in 2006 compared to the 506,000 in the first year of the Initiative (Partnership). For 2008–2010, the measles endemic countries of Southeast Asia are the targets of continuing campaigns.

Fortunately, measles virus has remained a monotypic agent with remarkably stable surface proteins that are responsible for induction of immunity. Forty-five years after introduction of the vaccine in 1963, it continues to provide solid, enduring immunity to vaccine recipients today, neutralizing measles viruses of all lineages. Even in those areas where exposure to wild measles viruses have been absent for many years, antibodies and resultant protection have persisted. An attack of natural measles conferred lifelong immunity to those who acquired it. Although it is tempting to predict that successful vaccination with attenuated measles virus will provide equivalent immunity, it is premature to make such a prediction with the passage of less than five decades since its initial availability. In an era where many individuals are living to their eighties and nineties, the senescence of their immune systems may not maintain what has been assumed to be lifelong immunity. Only by continuing longitudinal studies will the answer to this question be provided.

In September 1985, at age 88, John Enders died peacefully at his home while reading poetry. His vision of a measles-free world has come closer to reality than he anticipated, but the challenges of elimination and eradication of so highly transmissible a virus will continue to confront us for many more years. His legacy, however, endures without challenge.

References

Annunziato D, Kaplan MH, Hall WW, Ichinose H, Lin JH, Balsam D, Paladino VS (1982) Atypical measles syndrome: pathologic and serologic findings. Pediatrics 70:203–209

Bellini WJ, Rota JS, Lowe LE (2005) Subacute sclerosing panencephalitis: more cases of this diseases are prevented by measles immunization than was previously recognized. J Infect Dis 192:1684–1693

de Quadros CA, Izurieta H, Venczel L, Carrasco P (2004) Measles eradication in the Americas: progress to date. J Infect Dis (Suppl 1) 189:S227–S235

Enders JF, Peebles TC (1954) Propagation in tissue culture of cytopathogenic agents from patients with measles. Proc Soc Exp Biol Med 86:277–286

Enders JF, Katz SL, Milovanovic MV, Holloway A (1960) Studies on an attenuated measles-virus vaccine. I. Development and preparation of the vaccine: technics for assay of effects of vaccination. New Eng J Med 263:153–159

Enders JF, Katz SL, Holloway A (1962) Development of attenuated measles virus vaccines. A summary of recent investigations. Am J Dis Child 103:335–340

Fulginiti VA, Eller JJ, Downie AW, Kempe CH (1967) Altered reactivity to measles virus: atypical measles in children previously immunized with inactivated measles virus vaccines. JAMA 1075–1080

Horta-Barbosa L, Fucillo DA, Sever JL, Zevan W (1969) Subacute sclerosing panencephalitis: isolation of measles virus from a brain biopsy. Nature 221:974

Katz SL (2005) A vaccine-preventable disease kills half a million children annually. J Infect Dis 192:1679–1680

Katz SL, Enders JF (1959) Immunization of children with a live attenuated measles virus. Am J Dis Child 98:605–607

Katz SL, Hinman AR (2004) Summary and conclusions measles elimination meeting, 16–17 March 2000. J Infect Dis (Suppl 1) 189:S43–S47

Katz SL, Milovanovic MV, Enders JF (1958) Propagation of measles virus in cultures of chick embryo cells. Proc Soc Exp Biol Med 97:23–29

Katz Sl, Enders JF, Holloway A (1960a) Studies on an attenuated measles-virus vaccine. Clinical, virologic and immunologic effects of vaccine in institutionalized children. New Eng J Med 263:159–161

Katz SL, Kempe CH, Black FL, Lepow ML, Krugman S, Haggerty RJ, Enders JF (1960b) Studies on an attenuated measles-virus vaccine. VIII. General summary and evaluation of the results of vaccination. New Eng J Med 263:180–184

Katz SL, Morley DC, Krugman S (1962) Attenuated measles virus vaccine in Nigerian children. Am J Dis Child 103:402–405

Milovanovic MV, Enders JF, Mitus A (1957) Cultivation of measles virus in human amnion cells and developing chick embryo. Proc Soc Exp Biol Med 95:120–127

Payne FE, Baublis JV, Itahashi HH (1969) Isolation of measles virus from cell cultures of brain from a patient with subacute sclerosing panencephalitis. New Eng J Med 281:585–589

Peebles T, McCarthy K, Enders JF, Holloway A (1957) Behavior of monkeys after inoculation of virus derived from patients with measles and propagated in tissue culture. J Immunol 78:63–74

Rauh LW, Schmidt R (1965) Measles immunization with killed virus vaccine. Am J Dis Child 109:232–237

Wolfson LJ, Strebel PM, Gacic-Dobo M, Hoekstra EJ, McFarland JW, Hersh BS (2007) Has the 2005 measles mortality reduction goal been achieved? Lancet 369:191–200

Chapter 2
Measles Virus Receptors

Y. Yanagi(✉), M. Takeda, S. Ohno, and T. Hashiguchi

Contents

Introduction	14
Overview of MV Receptors	14
MV Receptors	16
CD46	16
SLAM	17
A Putative Receptor on Epithelial Cells	19
Other Receptors	20
Morbillivirus Receptors	21
Interaction of the MV H Protein with Receptors	21
MV Tropism and Pathogenesis in Relation to Receptors	24
Conclusions	25
References	26

Abstract Measles virus (MV) has two envelope glycoproteins, the hemagglutinin (H) and fusion protein, which are responsible for attachment and membrane fusion, respectively. Signaling lymphocyte activation molecule (SLAM, also called CD150), a membrane glycoprotein expressed on immune cells, acts as the principal cellular receptor for MV, accounting for its lymphotropism and immunosuppressive nature. MV also infects polarized epithelial cells via an as yet unknown receptor molecule, thereby presumably facilitating transmission via aerosol droplets. Vaccine and laboratory-adapted strains of MV use ubiquitously expressed CD46 as an alternate receptor through amino acid substitutions in the H protein. The crystal structure of the H protein indicates that the putative binding sites for SLAM, CD46, and the epithelial cell receptor are strategically located in different positions of the H protein. Other molecules have also been implicated in MV infection, although their relevance remains to be determined. The identification of MV receptors has advanced our understanding of MV tropism and pathogenesis.

Y. Yanagi
Department of Virology, Faculty of Medicine, Kyushu University, 812-8582, Fukuoka, Japan,
e-mail: yyanagi@virology.med.kyushu-u.ac.jp

Introduction

Measles virus (MV), a member of the genus *Morbillivirus* in the family *Paramyxoviridae*, is an enveloped virus with a nonsegmented negative-sense RNA genome (Griffin 2007). It has two envelope glycoproteins, the hemagglutinin (H) and fusion (F) protein, which are involved in virus entry. MV enters a cell by pH-independent membrane fusion at the cell surface. Attachment of the H protein to a cell surface receptor is thought to induce the conformational change of the H protein, which in turn activates the fusion activity of the adjacent F protein, resulting in the fusion of the viral envelope with the host cell membrane (see the chapter by C. Navaratnarajah et al., this volume, for a more detailed discussion). Upon infection of susceptible cells, MV usually causes cell–cell fusion, producing syncytia.

Some molecules (called entry receptors) are, by themselves, capable of inducing the conformational changes of the H and F proteins required for membrane fusion and thus allowing MV entry, whereas others (called attachment receptors) only allow MV attachment to the cell without the ensuing membrane fusion and entry. While entry receptors are indispensable for entry, attachment receptors may also play a significant role in MV infection of certain cells by increasing the entry efficiency. The presence of these viral receptors determines whether a specific cell type is susceptible to MV. However, for successful MV infection, the cell also has to be permissive for viral replication at post-entry steps, which depends on other intracellular host factors.

MV was first isolated using the primary culture of human kidney cells (Enders and Peebles 1954). This first isolate is the progenitor of currently used live vaccines of the Edmonston lineage (see the chapter by S.L. Katz and D. Griffin and C.-H. Pan, this volume). In the past, Vero cells derived from the African green monkey kidney were widely used to isolate MV from measles patients. However, the isolation with Vero cells was inefficient and usually required several blind passages. Then, it was demonstrated that the Epstein-Barr (EB) virus-transformed marmoset B lymphoblastoid cell line B95-8 and its subline B95a are highly susceptible to MV, and that B95a cell-isolated MV strains retain pathogenicity to experimentally infected monkeys, unlike Vero cell-isolated strains (Kobune et al. 1990, 1996). Thus, B95a became a standard cell line used for MV isolation, together with other human B cell lines (Lecouturier et al. 1996; Schneider-Schaulies et al. 1995). These developments set the stage for the identification of MV receptors.

Overview of MV Receptors

Initial attempts to identify an MV receptor employed the commonly used laboratory-adapted MV strains of the Edmonston lineage. Naniche et al. (1992) generated a monoclonal antibody (mAb) that was capable of inhibiting the cell–cell fusion induced by the H and F proteins of the Hallé strain of MV. The mAb precipitated a cell-surface glycoprotein, which was subsequently identified as CD46 (also called membrane cofactor protein) (Naniche et al. 1993). Transfection of the human CD46 gene conferred susceptibility to MV on resistant rodent cell lines. Dörig et al. (1993) independently showed that human CD46 acts as a receptor for the Edmonston strain of MV. CD46 is expressed on all nucleated human cells (see the chapter by C. Kemper and J.P. Atkinson, this volume), thus explaining the ability of these laboratory-adapted MV strains to grow well in most human cell lines. In monkeys, CD46 is also present on red blood cells, consistent with the observation that these strains hemagglutinate monkey red blood cells.

Unlike vaccine and laboratory-adapted strains, MV strains isolated in B95a or human B cell lines were found to grow only in some lymphoid cell lines (Kobune et al. 1990; Schneider-Schaulies et al. 1995; Tatsuo et al. 2000a). Furthermore, the H protein from B cell line-isolated MV strains neither induced downregulation of CD46 nor caused cell–cell fusion (upon co-expression of the F protein) in CD46-positive cell lines (Bartz et al. 1998; Lecouturier et al. 1996; Tanaka et al. 1998). These observations suggested that B cell line-isolated MV strains may utilize a molecule other than CD46 as a receptor (Bartz et al. 1998; Buckland and Wild 1997; Hsu et al. 1998; Lecouturier et al. 1996; Tanaka et al. 1998; Tatsuo et al. 2000a). Using an expression cloning approach, Tatsuo et al. (2000b) isolated a cDNA that could render a resistant cell line susceptible to B95a cell-isolated MV strains. The cDNA clone encoded signaling lymphocyte activation molecule (SLAM, also called CD150), a membrane glycoprotein expressed on immune cells (Cocks et al. 1995). Importantly, the Edmonston strain was found to use SLAM, in addition to CD46, as a receptor, indicating that SLAM acts as a receptor not only for B cell line-isolated MV strains but also vaccine and laboratory-adapted strains (Tatsuo et al. 2000b). Other groups have reached the same conclusions using different approaches (Erlenhoefer et al. 2001; Hsu et al. 2001).

To determine the receptor usage of different MV strains, Erlenhöfer et al. (2002) examined a number of MV strains with various isolation and passage histories, and showed that SLAM acts as a common receptor for all MV strains tested. In fact, no MV strain that does not use SLAM as a receptor has ever been reported, except artificially generated SLAM-blind recombinant viruses (Vongpunsawad et al. 2004). In general, B cell line-isolated strains utilize SLAM but not CD46 as a receptor, whereas vaccine and Vero cell-isolated strains use both SLAM and CD46 as receptors. In an attempt to determine differential receptor usage in vivo, viruses in throat swabs from measles patients were plaque-titrated on Vero cells with or without human SLAM expression. The results showed that like B cell line-isolated MV strains, the great majority of viruses in vivo use SLAM but not CD46 as a receptor (Ono et al. 2001a). Manchester et al. (2000) reported that clinical isolates obtained in peripheral blood mononuclear cells (PBMCs) utilize CD46 as a receptor. However, these strains replicated well in Chinese hamster ovary (CHO) cells expressing human SLAM (Tatsuo et al. 2000b) but failed to productively infect CHO cells expressing human CD46 (Manchester et al. 2000). Taken together, the data indicate that SLAM acts as the principal cellular receptor for MV in vivo, and that the use of CD46 may be the result of MV adaptation in vitro.

In addition to these two well-characterized receptors, there may be other molecules that can act as an MV receptor. Pathological studies with humans and experimentally infected monkeys have shown that MV infects not only immune cells, but also epithelial, endothelial, and neuronal cells (Griffin 2007), all of which do not express SLAM. Now there is good evidence for the presence of an MV receptor on polarized epithelial cells (Tahara et al. 2008; Takeda et al. 2007). Furthermore, a ubiquitously expressed molecule(s) has been shown to allow MV infection, but not syncytium formation, in various types of cells from many species at low efficiencies (Hashimoto et al. 2002).

Figure 1 summarizes the receptor usage of MV in vivo and in vitro. In the following section, the properties of MV receptors will be discussed individually, along with their roles in MV infection.

MV Receptors

CD46

CD46 acts as a receptor for vaccine and laboratory-adapted strains of MV. Its physiological function is to protect cells from attack by autologous complement, by regulating

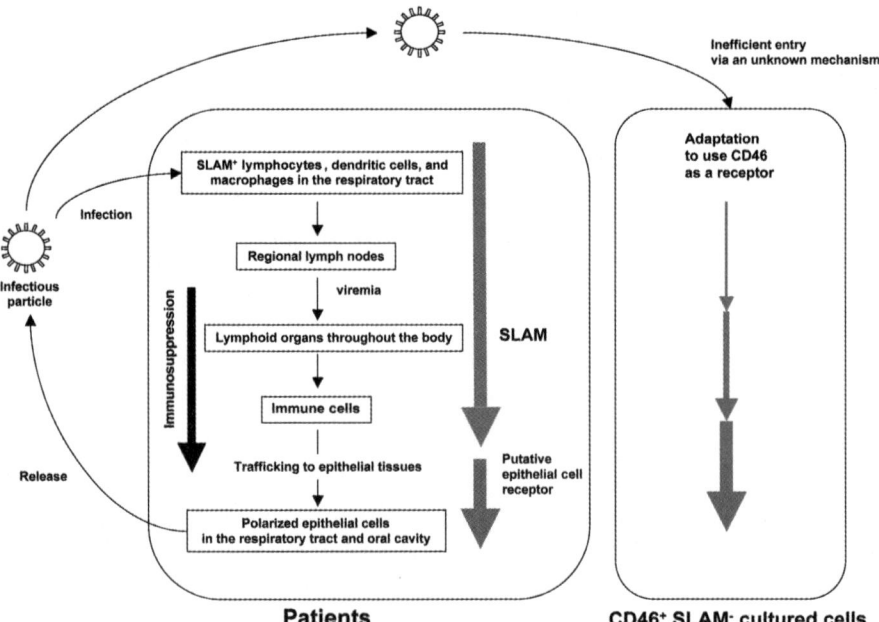

Fig. 1 Receptor usage of MV in vivo and in vitro. SLAM acts as the principal receptor for MV in vivo, accounting for its lymphotropism and immunosuppressive nature. MV also infects polarized epithelial cells via an as yet unknown receptor molecule, releasing progeny infectious particles for transmission. Vaccine and laboratory-adapted strains of MV acquire the ability to use CD46 as an alternate receptor through amino acid substitutions in the H protein during in vitro culture

complement activation. Furthermore, CD46 signaling has been implicated in the regulation of innate and acquired immune responses. The structure and functions of human CD46 are reviewed in the chapter by C. Kemper and J.P. Atkinson, this volume.

The ectodomain of the H protein of the Edmonston strain binds to the most membrane-distal short consensus repeat 1 and 2 of CD46 (Devaux et al. 1996; Iwata et al. 1995; Manchester et al. 1995). Analyses of the H proteins from many MV strains have revealed that the majority of strains using both SLAM and CD46 as receptors have tyrosine at position 481, whereas most B cell line-isolated strains have asparagine at that position. An N481Y substitution in the H protein was shown to allow B cell line-isolated strains to use CD46 as a receptor, without affecting their ability to use SLAM (Bartz et al. 1996; Erlenhöfer et al. 2002; Hsu et al. 1998; Lecouturier et al. 1996; Nielsen et al. 2001; Shibahara et al. 1994; Xie et al. 1999). Furthermore, when B cell line-isolated strains were adapted to growth in Vero cells (SLAM-negative), an N481Y substitution of the H protein was often observed after several passages (Nielsen et al. 2001; Schneider et al. 2002; Shibahara et al. 1994). Some Vero cell-adapted strains have a serine to glycine substitution at position 546 of the H protein, instead of the N481Y substitution (Li and Qi 2002; Rima et al. 1997; Shibahara et al. 1994; Woelk et al. 2001).

Is a single N481Y or S546G substitution in the H protein sufficient for MV to use CD46 as a receptor? By using recombinant viruses, it was shown that an N481Y or S546G substitution in the H protein alone cannot make a B cell line-isolated MV strain utilize CD46 as efficiently as the Edmonston strain (Seki et al. 2006). Several additional mutations are required for the H protein to interact efficiently with CD46 (Tahara et al. 2007a). This may explain why CD46-using viruses are seldom detected in vivo (Ono et al. 2001a). Furthermore, CD46-using viruses may have a growth disadvantage because they induce higher levels of type I interferons in PBMCs (Naniche et al. 2000). Thus, CD46-using viruses may emerge and grow in SLAM-negative cultured cells, but they may not expand in vivo because there is little selection pressure for them (the interferon system may even act against them). Although CD46 is the first MV receptor identified, its relevance in vivo remains to be proven.

SLAM

It is now well established that SLAM is the principal cellular receptor for MV. SLAM is a member of the SLAM family receptors that mediate important regulatory signals in immune cells (reviewed in Engel et al. 2003; Ma et al. 2007; Sidorenko and Clark 2003; Veillette 2006). SLAM is expressed on thymocytes, activated lymphocytes, mature dendritic cells (DCs), macrophages, and platelets in both humans and mice. SLAM is not expressed on monocytes (see below in this section for activated monocytes), natural killer cells, or granulocytes. SLAM has two extracellular immunoglobulin superfamily domains, V and C2, and a cytoplasmic tail with three tyrosine-based motifs that undergo phosphorylation and recruit SH2 domain-containing proteins such as SLAM-associated protein (SAP) and

Ewing's sarcoma-associated transcript 2 (EAT-2) (Cocks et al. 1995; Engel et al. 2003; Ma et al. 2007; Veillette 2006) (Fig. 2). SLAM functions by interacting with another SLAM molecule present on an adjacent cell (Mavaddat et al. 2000). Ligation of SLAM on CD4⁺ T cells leads to its binding to SAP, which in turn recruits and activates the Src-related protein tyrosine kinase FynT, resulting in tyrosine phosphorylation of SLAM. Combined with T cell receptor engagement, this triggers downstream effectors, leading to upregulation of the GATA-3 transcription factor and production of T helper 2 cytokines such as interleukin (IL)-4 and IL-13 (Engel et al. 2003; Ma et al. 2007; Veillette 2006) (Fig. 2). SLAM also regulates lipopolysaccharide-induced production of IL-12, tumor necrosis factor α, and nitric oxide by macrophages in mice (Wang et al. 2004). The distribution and functions of SLAM provide a good explanation for the lymphotropism and immunosuppressive nature of MV. Indeed, a recent study of MV infection in macaques identified SLAM⁺ lymphocytes and DCs as predominantly infected cell types (de Swart et al. 2007). Although SLAM is reported to be a marker for the

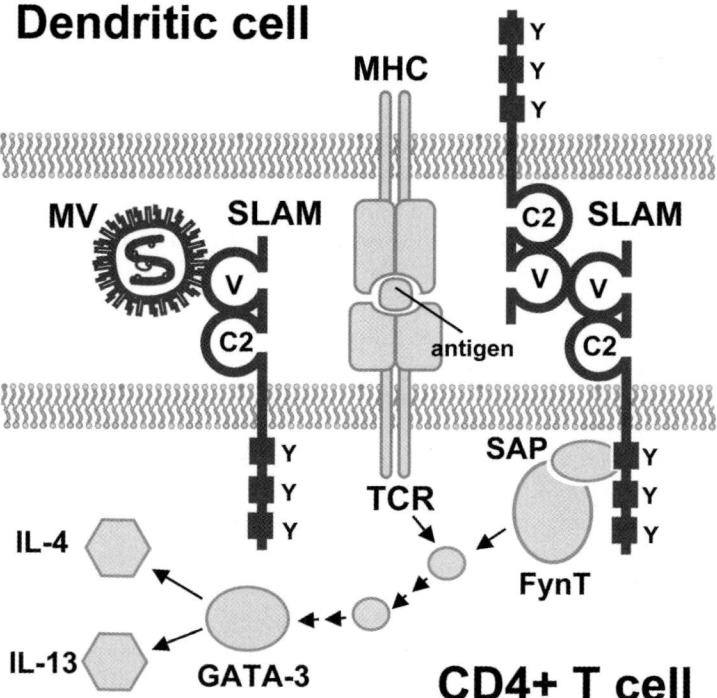

Fig. 2 SLAM structure and signal transduction. SLAM has extracellular V and C2 domains and a cytoplasmic tail with tyrosine (Y)-based motifs. Its ligand is another SLAM present on adjacent cells. Ligation of SLAM on CD4⁺ T cells leads to its binding to SAP, which in turn recruits and activates FynT, resulting in phosphorylation of SLAM. Combined with T cell receptor (TCR)-mediated signals, this triggers downstream effectors, leading to upregulation of GATA-3 and production of IL-4 and IL-13. The MV H protein binds to the V domain of SLAM to initiate cell entry. *MHC*, major histocompatibility complex

most primitive hematopoietic stem cells in mice (Kiel et al. 2005), it is currently unknown whether it is also expressed on human hematopoietic stem cells, thereby contributing to MV pathogenesis.

Mouse SLAM has functional and structural similarity to human SLAM (~60% identity at the amino acid level), but it cannot act as a receptor for MV, partly explaining why mice are not susceptible to MV (Ono et al. 2001b). The V domain of human SLAM is necessary and sufficient for MV receptor function (Ono et al. 2001b) (Fig. 2), and the amino acid residues at positions 60, 61, and 63 are critical for function (Ohno et al. 2003). Substitutions at these three positions to human-type residues make mouse SLAM act as an MV receptor, while introduction of changes at these positions compromises the receptor function of human SLAM. At present, it is unknown whether these residues directly bind to the H protein or, upon substitutions, modulate the conformation of SLAM, thereby affecting its interaction with the H protein. The answer to this question awaits the elucidation of the crystal structure of human SLAM complexed with the MV H protein.

Toll-like receptor (TLR) 2, 4, and 5 ligands induce SLAM expression on monocytes (Bieback et al. 2002; Farina et al. 2004; Minagawa et al. 2001). The MV H protein also induces SLAM expression on monocytes after binding to TLR 2 (Bieback et al. 2002). Thus, MV may induce its own entry receptor on potential target cells such as TLR2$^+$ monocytes and DCs.

Vero cells stably expressing human SLAM (Vero/hSLAM) (Ono et al. 2001a) are now commonly used for MV isolation and propagation, replacing EB virus-producing B95a cells for safety reasons. However, in retrospect, the introduction of SLAM$^+$ B95a cells to MV research (Kobune et al. 1990) was critical for the identification of SLAM as the principal receptor for MV. Without the use of B95a cells, its identification would have been delayed for many years.

A Putative Receptor on Epithelial Cells

In measles patients and experimentally infected monkeys, MV antigens and syncytia have been identified in the epithelia of various organs, including the skin, oral cavity, pharynx, trachea, esophagus, intestines, and urinary bladder, as well as in lymphoid tissues (for references, see the paper by Takeda et al. 2007). However, epithelial cells do not express SLAM, and B cell line-isolated MV strains, unlike vaccine and laboratory-adapted strains, do not infect most epithelial cell lines. Takeuchi et al. (2003) reported that a B95a cell-isolated strain caused syncytium formation in primary human respiratory epithelial cells, which was independent of SLAM and CD46. Recently, a human lung adenocarcinoma cell line NCI-H358 (Takeda et al. 2007) as well as four human polarized epithelial cell lines (Tahara et al. 2008) were shown to support SLAM- and CD46-independent MV entry, replication, and syncytium formation. Furthermore, analyses using anti-H protein mAbs and recombinant viruses possessing the mutated H proteins indicated that the receptor-binding site on the H protein required to infect these epithelial cell lines is different from the binding sites for SLAM and CD46 (see p. 22). Thus, wild-type viruses circulating in measles patients appear to have an intrinsic ability to infect immune and polarized epithelial cells by using SLAM and an as yet unidentified molecule, respectively. (Also see the chapter by C. Navaratnarajah et al., this volume.)

Other Receptors

The C-type lectin DC-specific intercellular adhesion molecule 3-grabbing nonintegrin (DC-SIGN) may play a role in MV infection of DCs (de Witte et al. 2006). Both attachment and infection of DCs with MV are blocked in the presence of DC-SIGN inhibitors. However, stable expression of DC-SIGN cannot confer susceptibility to MV on CHO cells. Thus, DC-SIGN appears to act as an attachment, but not entry, receptor for DCs.

On rare occasions, MV causes subacute sclerosing panencephalitis (SSPE), a persistent MV infection in the central nervous system (CNS), in which few free MV particles and syncytia are detected (see the chapters by V.A. Young and G. Rall and M.B. Oldstone, this volume). MV may spread trans-synaptically in neurons (Lawrence et al. 2000). It has been proposed, based on competitive inhibition studies and on experiments with knock-out mice, that neurokinin-1 (NK-1, substance P receptor) may promote MV entry into neurons by serving as a receptor for the MV F protein (Makhortova et al. 2007). However, the exact mechanism by which NK-1 contributes to MV spread in neurons remains to be defined.

Studies with recombinant MVs expressing green fluorescent protein (GFP) demonstrated that SLAM- and CD46-independent MV entry occurs in a variety of cell lines (Hashimoto et al. 2002). This mode of entry produces solitary infected cells, but does not usually induce syncytium formation, and its efficiency is 100- to 1,000-fold lower than that of SLAM-dependent entry. Such a weak MV-receptor interaction that only allows inefficient entry may not lead to apparent cell–cell fusion (Hasegawa et al. 2007). This inefficient entry appears to be mediated by a ubiquitously expressed molecule(s) because it occurs in almost all cultured cells from various species (Hashimoto et al. 2002). It has been reported that B cell line-isolated MV strains effectively infect human umbilical vein and brain microvascular endothelial cells (Andres et al. 2003). Shingai et al. (2003) showed that pseudotype viruses bearing the H and F proteins of SSPE strains of MV utilize SLAM, but not CD46, as a receptor, and that they can infect various SLAM-negative cell lines independent of CD46. It remains to be determined whether MV infection of endothelial and neuronal cells in these instances is mediated by the ubiquitous inefficient receptor or a more efficient receptor(s) such as the putative epithelial cell receptor.

Even this ubiquitous inefficient receptor may allow significant MV growth after virus adaptation to cultured cells at post-entry step(s) of the viral life cycle. Recombinant chimeric viruses were generated, in which part of the genome of a B cell line-isolated MV strain was replaced with the corresponding genes from the Edmonston strain. While the parental virus could not grow in SLAM-negative Vero cells, the virus possessing the Edmonston H gene replicated efficiently using CD46 as a receptor. The recombinant virus possessing the Edmonston M or L gene also grew in Vero cells, although their entry efficiencies were as low as that of the parental virus (Tahara et al. 2005). This study provides an explanation for the previous observations that the recombinant viruses based on the Edmonston strain possessing the H protein of B cell line-isolated strains efficiently replicate in Vero cells (Johnston et al. 1999; Takeuchi et al. 2002). Other studies have also shown that B cell line-isolated MV strains can adapt to growth in Vero cells by substitutions in other proteins than the receptor-binding H protein (Bankamp et al. 2008; Kouomou and Wild 2002; Miyajima et al. 2004; Takeuchi et al. 2000). The changes found in these proteins may enhance MV growth at post-entry step(s) by improving viral transcription and replication, virus assembly (Tahara et al. 2007b), and/or evasion of antiviral host responses, thereby compensating the inefficient entry.

Morbillivirus Receptors

MV is a member of the *Morbillivirus* genus, which also includes canine distemper virus (CDV), rinderpest virus (RPV), peste-des-petits-ruminants virus, cetacean morbillivirus, and phocine distemper virus (Griffin 2007). Morbilliviruses are lymphotropic and cause lymphopenia and immunosuppression in respective host species. The common tropism and pathology of these viruses prompted Tatsuo et al. (2001) to examine the receptor usage of several strains of CDV and RPV. That study showed that all CDV and RPV strains examined use dog and cow SLAM as a receptor, respectively.

Dog and ferret macrophages (Appel and Jones 1967; Poste 1971), mitogen-stimulated dog lymphocytes (Appel et al. 1992), and the marmoset B cell line B95a (Kai et al. 1993) have been successfully used to isolate virulent CDV. All these cells presumably express SLAM. Moreover, CDV was readily isolated in Vero cells stably expressing dog SLAM (Vero.DogSLAMtag) from the majority of dogs with distemper, suggesting that CDV uses dog SLAM as the principal receptor in vivo (Seki et al. 2003). This is supported by the finding that a recombinant CDV unable to recognize SLAM is attenuated in experimental infection of ferrets (von Messling et al. 2006). It is currently unknown how CDV infects the cells in the CNS, one of the commonly affected targets. It was shown that a wild-type RPV uses cow SLAM as a receptor, while the Plowright vaccine strain of RPV can use heparan sulphate as an alternative receptor, growing in many types of cells (Baron 2005). Thus, the use of SLAM as a receptor may be a common property of all morbilliviruses.

Interaction of the MV H protein with Receptors

Binding of CD46 and SLAM to the MV H protein has been studied using soluble molecules (Hashiguchi et al. 2007; Navaratnarajah et al. 2008; Santiago et al. 2002). SLAM binds to the MV Edmonston (vaccine strain) and IC-B (B95a cell-isolated wild-type strain) H proteins with similar affinities (dissociation constant Kd of 0.43 vs 0.29 µM). On the other hand, CD46 binds to the Edmonston H protein (Kd of 2.2 µM) but not to the IC-B H protein (Hashiguchi et al. 2007).

To identify residues in the MV H protein involved in the interaction with SLAM and CD46, a series of mutants of the Edmonston or Hallé H protein were examined for their ability to induce SLAM- or CD46-dependent cell–cell fusion or to downregulate SLAM or CD46 from the cell surface (Massé et al. 2002, 2004; Navaratnarajah et al. 2008; Vongpunsawad et al. 2004). Changes of the relevant residues are expected to affect these functions of the H protein. The studies showed that some residues (I194, D505, D507, Y529, D530, T531, R533, H536, F552, Y553, and P554) interact with SLAM, and others (A428, F431, V451, Y452, L464, Y481, P486, I487, A527, S546, S548, and F549) interact with CD46. Tahara et al. (2007a) showed that substitutions at positions 390, 416, 446, 484, and 492, in addition to N481Y, are important to allow the IC-B strain to use CD46 as a receptor, suggesting that amino acid residues at those positions may also interact with CD46. Using site-directed mutagenesis, it was recently shown that aromatic residues such as F483, Y541, and Y543 in the H protein are critical for MV to infect and cause cell–cell fusion in polarized epithelial cell lines (Tahara et al. 2008). (Also see the chapter by C. Navaratnarajah et al., this volume.)

The crystal structure of the MV H protein was recently determined (Hashiguchi et al. 2007). The receptor-binding head domain forms a disulfide-linked homodimer and exhibits a six-bladed β-propeller fold (β1–β6). The residues implicated in the interaction with SLAM, CD46 or the putative epithelial cell receptor are indicated on the determined crystal structure of the MV H protein (Fig. 3A, viewed from the top of the monomer). SLAM-relevant residues are mapped to the interstrand loops of the β5 sheet. The key residues for the interaction with CD46 span the β3–β5 sheets of the side face of the head domain and are mapped in different locations from the putative SLAM-binding site. The aromatic residues implicated in the interaction with the putative epithelial cell receptor are located between the putative SLAM- and CD46-binding sites. Notably, the residues implicated in the interaction with SLAM or the putative epithelial cell receptor are highly conserved among morbilliviruses, whereas those shown to be important for the interaction with CD46 are not. Thus, it is likely that many morbilliviruses, including MV, CDV and RPV, use their orthologs (SLAM and an unknown molecule) to infect immune and epithelial cells, respectively. Importantly, residues relevant for the interaction with SLAM and the putative epithelial cell receptor are located upward from the viral envelope, because of the tilted orientation of the molecules forming the H protein dimer (Fig. 3B, 3C). Thus, they may readily interact with SLAM on immune cells and the putative receptor on epithelial cells. On the other hand, CD46-relevant residues are accessible from the top of the H protein, but are located more to the side. Although most of these residues are expected to interact directly with the respective receptors because of their location on the surface of the H protein, elucidation of the crystal structures

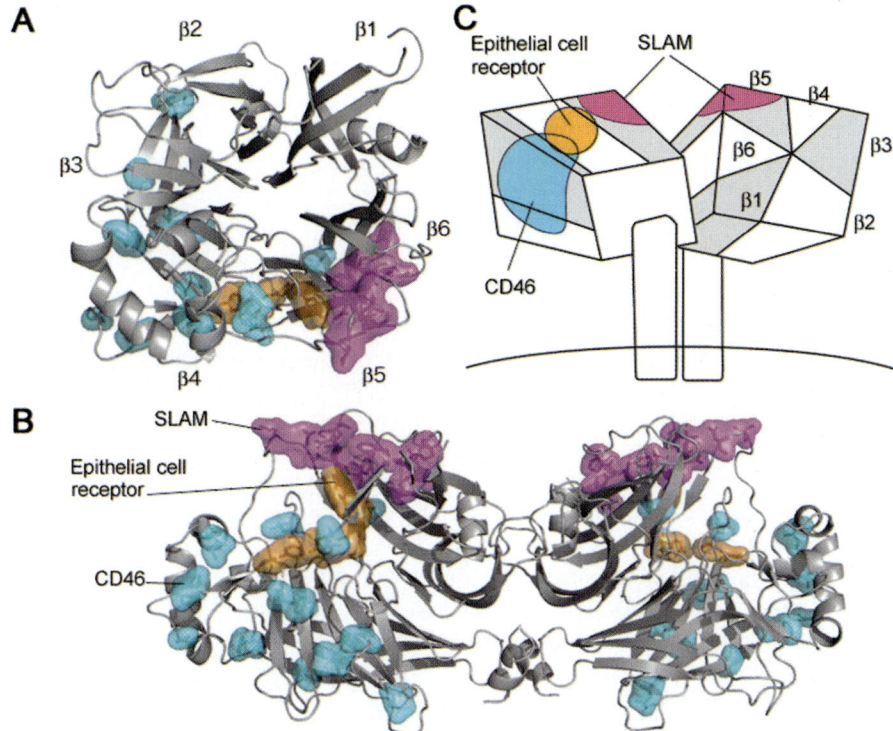

Fig. 3 A–C Receptor-binding sites on the MV H protein. The receptor-binding head domain of the MV H protein comprises six β-sheets arranged cyclically around an axis as the blades of a propeller, and forms a homodimer (**A**) The head domain monomer of the H protein viewed from the top of the propeller-like structure (with the axis in the center) is shown by the ribbon model, together with residues implicated in the interaction with SLAM (*magenta*), CD46 (*cyan*), and a putative epithelial cell receptor (*orange*) (**B**) The H protein homodimer viewed from the side is shown by the ribbon model, with residues implicated in the interaction with receptors (**C**) The cartoon model of the H protein homodimer on the MV envelope. Residues implicated in the interaction with SLAM and a putative epithelial cell receptor are located upward from the viral envelope because of the tilted orientation of the molecules forming the homodimer

of the MV H protein complexed with individual receptors are required to determine whether they indeed bind to the receptors.

It has been suggested that SLAM and CD46 bind to overlapping sites in the H protein of the Edmonston strain, based on receptor binding competition and on blocking of SLAM and CD46 binding with the same anti-H protein mAbs (Santiago et al. 2002). However, given the assumed locations of the respective receptor-binding sites, it is more likely that the observed competition and blocking occurred because of a mechanism of steric hindrance. Similarly, many mAbs neutralizing SLAM-dependent MV infection appear to do so by steric hindrance, because their mapped epitopes are located in different positions from the putative SLAM-binding sites (Bouche et al. 2002; Hashiguchi et al. 2007; Santibanez et al. 2005).

MV Tropism and Pathogenesis in Relation to Receptors

Identification of MV receptors has led to better understanding of MV tropism and pathogenesis. Lymphotropism of MV is explained by infection of SLAM⁺ immune cells. Polarized epithelial cells appear to express a specific cellular receptor for MV. Currently, it is not known how endothelial and neuronal cells are infected by MV. They may have their own receptors or express the same receptor molecule as epithelial cells. Alternatively, they may be infected by MV via an inefficient receptor.

MV is transmitted via aerosol droplets. Although respiratory epithelial cells are generally suspected, initial target cells are not well defined. A classical study on CDV infection of dogs reported that the virus was detected only in bronchial lymph nodes and in tonsils on the day of infection, and that it appeared in mononuclear cells of the blood on the 2nd and 3rd days (Appel 1969). A ferret model of CDV infection also showed massive lymphocyte infection in PBMCs and lymphoid organs including the thymus, spleen, and lymph nodes, followed by infection of epithelial cells during the later stages of infection (von Messling et al. 2003, 2004). In intratracheal infection of macaques with GFP-expressing recombinant wild-type MV, de Swart et al. (2007) demonstrated that SLAM⁺ lymphocytes and DCs are predominantly infected cell types, with an occasional infection of epithelial cells at the later stage. Thus, it is likely that the primary targets of MV are SLAM⁺ immune cells in the respiratory tract, such as lymphocytes, DCs, and macrophages, rather than epithelial cells. It is also possible that DC-SIGN⁺ DCs capture MV (without being infected) in the respiratory epithelia and carry it to local lymph nodes, where the virus is transferred to activated (SLAM⁺) lymphocytes (de Swart et al. 2007; de Witte et al. 2006, 2008).

These interpretations are consistent with the observation that MV infects SLAM⁺ immune cells more efficiently than it does polarized epithelial cells (M. Takeda et al., unpublished observations). Most likely, at the later stage of infection when a large amount of MV is produced, infected SLAM⁺ immune cells may transfer the virus, albeit inefficiently, to epithelial cells, which in turn propagate the virus via the epithelial cell receptor. Since polarized epithelial cells with tight junctions cover the external epithelial surface, MV may not efficiently release progeny virus particles into the external surface through its ability to infect SLAM⁺ immune cells alone. Furthermore, studies showed that MV is selectively released into the apical (luminal) side of polarized epithelial cells (Tahara et al. 2008). Thus, the ability to infect epithelial cells may be necessary for MV to spread efficiently from person to person. This may also explain why MV is transmitted efficiently via aerosol droplets, whereas human immunodeficiency virus (HIV), which shares the tropism for immune cells with MV, is transmitted exclusively via sexual contact or blood.

Two clinical observations are of particular interest, which may be understood in terms of the use of SLAM as a receptor by MV. First, Burkitt's lymphoma and Hodgkin's disease have been reported to regress after MV infection (Bluming and Ziegler 1971; Taqi et al. 1981). EB virus may be responsible for these diseases, and

EB virus-transformed B lymphoid cell lines have been shown to express high levels of SLAM (Aversa et al. 1997; Tatsuo et al. 2000b). Thus, it is likely that these EB virus-related tumors expressed SLAM, and MV infected and killed these tumor cells. Second, HIV replication is reported to be suppressed during acute measles (Moss et al. 2002). Although the authors propose that the finding is related to the ability of MV to suppress lymphocyte proliferation (Garcia et al. 2005), it is also likely that suppression of HIV replication occurs because MV targets the very cells that harbor HIV provirus and allow HIV replication. HIV resides and replicates in memory and activated $CD4^+$ T cells, which are likely to be $SLAM^+$, and therefore to be infected and killed by MV. A similar mechanism may also explain in part why measles is more severe among people in developing countries, where chronic infection with various pathogens may increase the percentage of activated lymphocytes, which are $SLAM^+$ and susceptible to MV infection.

Immunosuppression and lymphopenia are characteristic of measles. Infection and subsequent destruction of $SLAM^+$ immune cells may account for these immunological abnormalities. Furthermore, MV infection may also affect SLAM signal transduction of immune cells by mimicking the natural ligand, thereby leading to the immune dysfunction (see the chapter by D. Gerlier and H. Valentin, this volume). MV-induced immunosuppression is discussed in more detail in the chapter by S. Schneider-Schaulies and J. Schneider-Schaulies, this volume.

Conclusions

Although the identification of SLAM as the principal cellular receptor for MV has provided insight into MV tropism and pathogenesis, many problems associated with measles still remain to be clarified. In this regard, animal models such as macaques (see the chapter by R. de Swart et al., this volume) and human SLAM-expressing mice (see the chapter by C.I. Sellin and B. Horvat et al., this volume) are expected to provide useful information. For example, SLAM-knock-in mice have been shown to reproduce MV tropism and immunosuppression seen in human patients (Ohno et al. 2007; S. Ohno et al., unpublished observations). Identification of the epithelial cell receptor is greatly desired. The mechanism by which MV spreads in the CNS during SSPE is almost unknown. Further studies on these subjects, coupled with crystal structures of the MV H protein complexed with respective receptors, will lead to better understanding of MV pathogenesis and to novel strategies of the prevention and therapy of measles.

Acknowledgements We thank all members of our laboratory for helpful discussion. We also thank Drs. D. Gerlier and C. Navaratnarajah for their comments on the manuscript. Our work was supported by grants from the Ministry of Education, Culture, Sports, Science and Technology and the Ministry of Health, Labor and Welfare of Japan, and from Japan Society for the Promotion of Science.

References

Andres O, Obojes K, Kim KS, tel Meulen V, Schneider-Schaulies J (2003) CD46- and CD150-independent endothelial cell infection with wild-type measles viruses. J Gen Virol 84:1189–1197

Appel MJ (1969) Pathogenesis of canine distemper. Am J Vet Res 30:1167–1182

Appel MJG, Jones OR (1967) Use of alveolar macrophages for cultivation of canine distemper virus. Proc Soc Exp Biol Med 126:571–574

Appel MJ, Pearce-Kelling S, Summers BA (1992) Dog lymphocyte cultures facilitate the isolation and growth of virulent canine distemper virus. J Vet Diagn Invest 4:258–263

Aversa G, Chang C-C, Carballido JM, Cocks BG, de Vries JE (1997) Engagement of the signaling lymphocytic activation molecule (SLAM) on activated T cells results in IL-2-independent, cyclosporin A-sensitive T cell proliferation and IFN-gamma production. J Immunol 158:4036–4044

Bankamp B, Hodge G, McChesney MB, Bellini WJ, Rota PA (2008) Genetic changes that affect the virulence of measles virus in a rhesus macaque model. Virology 373:39–50

Baron MD (2005) Wild-type Rinderpest virus uses SLAM (CD150) as its receptor. J Gen Virol 86:1753–1757

Bartz R, Brinckmann U, Dunster LM, Rima B, ter Meulen V, Schneider-Schaulies J (1996) Mapping amino acids of the measles virus hemagglutinin responsible for receptor (CD46) downregulation. Virology 224:334–337

Bartz R, Firsching R, Rima B, ter Meulen V, Schneider-Schaulies J (1998) Differential receptor usage by measles virus strains. J Gen Virol 79:1015–1025

Bieback K, Lien E, Klagge I, Avota E, Schneider-Schaulies J, Duprex W, Wagner H, Kirschning C, ter Meulen V, Schneider-Schaulies S (2002) Hemagglutinin protein of wild-type measles virus activates toll-like receptor 2 signaling. J Virol 76:8729–8736

Bluming AZ, Ziegler JL (1971) Regression of Burkitt's lymphoma in association with measles infection. Lancet ii:105–106

Bouche FB, Ertl OT, Muller CP (2002) Neutralizing B cell response in measles. Viral Immunol 15:451–471

Buckland R, Wild TF (1997) Is CD46 the receptor for measles virus? Virus Res 48:1–9

Cocks BG, Chang C-CJ, Carballido JM, Yssel H, de Vries JE, Aversa G (1995) A novel receptor involved in T-cell activation. Nature 376:260–263

de Swart RL, Ludlow M, de Witte L, Yanagi Y, van Amerongen G, McQuaid S, Yuksel S, Geijtenbeek TB, Duprex WP, Osterhaus AD (2007) Predominant infection of CD150+ lymphocytes and dendritic cells during measles virus infection of macaques. PLoS Pathog 3:e178

de Witte L, Abt M, Schneider-Schaulies S, van Kooyk Y, Geijtenbeek TB (2006) Measles virus targets DC-SIGN to enhance dendritic cell infection. J Virol 80:3477–3486

de Witte L, de Vries RD, van der Vlist M, Yüksel S, Litjens M, de Swart RL, Geijtenbeek TBH (2008) DC-SIGN and CD150 have distinct roles in transmission of measles virus from dendritic cells to T-lymphocytes. PLoS Pathog 4:e1000049

Devaux P, Loveland B, Christiansen D, Milland J, Gerlier D (1996) Interactions between the ectodomains of haemagglutinin and CD46 as a primary step in measles virus entry. J Gen Virol 77:1477–1481

Dörig RE, Marcil A, Chopra A, Richardson CD (1993) The human CD46 molecule is a receptor for measles virus (Edmonston strain). Cell 75:295–305

Enders JF, Peebles TC (1954) Propagation in tissue cultures of cytopathic agents from patients with measles. Proc Soc Exp Biol Med 86:277–286

Engel P, Eck MJ, Terhorst C (2003) The SAP and SLAM families in immune responses and X-linked lymphoproliferative disease. Nat Rev Immunol 3:813–821

Erlenhoefer C, Wurzer WJ, Loffler S, Schneider-Schaulies S, ter Meulen V, Schneider-Schaulies J (2001) CD150 (SLAM) is a receptor for measles virus but is not involved in viral contact-mediated proliferation inhibition. J Virol 75:4499–4505

Erlenhöfer C, Duprex W, Rima B, ter Meulen V, Schneider-Schaulies J (2002) Analysis of receptor (CD46, CD150) usage by measles virus. J Gen Virol 83:1431–1436

Farina C, Theil D, Semlinger B, Hohlfeld R, Meinl E (2004) Distinct responses of monocytes to Toll-like receptor ligands and inflammatory cytokines. Int Immunol 16:799–809

Garcia M, Yu XF, Griffin DE, Moss WJ (2005) In vitro suppression of human immunodeficiency virus type 1 replication by measles virus. J Virol 79:9197–9205

Griffin DE (2007) Measles virus. In: DM Knipe, PM Howley, DE Griffin, RA Lamb, MA Martin, B Roizman, SE Straus (eds) Fields virology, 5th edn, Lippincott Williams & Wilkins, Philadelphia, pp 1551–1585

Hasegawa K, Hu C, Nakamura T, Marks JD, Russell SJ, Peng KW (2007) Affinity thresholds for membrane fusion triggering by viral glycoproteins. J Virol 81:13149–13157

Hashiguchi T, Kajikawa M, Maita N, Takeda M, Kuroki K, Sasaki K, Kohda D, Yanagi Y, Maenaka K (2007) Crystal structure of measles virus hemagglutinin provides insight into effective vaccines. Proc Natl Acad Sci U S A 104:19535–19540

Hashimoto K, Ono N, Tatsuo H, Minagawa H, Takeda M, Takeuchi K, Yanagi Y (2002) SLAM (CD150)-independent measles virus entry as revealed by recombinant virus expressing green fluorescent protein. J Virol 76:6743–6749

Hsu E, Iorio C, Sarangi F, Khine A, Richardson C (2001) CDw150 (SLAM) is a receptor for a lymphotropic strain of measles virus and may account for the immunosuppressive properties of this virus. Virology 279:9–21

Hsu EC, Sarangi F, Iorio C, Sidhu MS, Udem SA, Dillehay DL, Xu W, Rota PA, Bellini WJ, Richardson CD (1998) A single amino acid change in the hemagglutinin protein of measles virus determines its ability to bind CD46 and reveals another receptor on marmoset B cells. J Virol 72:2905–2916

Iwata K, Seya T, Yanagi Y, Pesando JM, Johnson PM, Okabe M, Ueda S, Ariga H, Nagasawa S (1995) Diversity of sites for measles virus binding and for inactivation of complement C3b and C4b on membrane cofactor protein CD46. J Biol Chem 270:15148–15152

Johnston ICD, ter Meulen V, Schneider-Schaulies J, Schneider-Schaulies S (1999) A recombinant measles vaccine virus expressing wild-type glycoproteins: consequences for viral spread and cell tropism. J Virol 73:6903–6915

Kai C, Ochikubo F, Okita M, Iinuma T, Mikami T, Kobune F, Yamanouchi K (1993) Use of B95a cells for isolation of canine distemper virus from clinical cases. J Vet Med Sci 55:1067–1070

Kiel MJ, Yilmaz OH, Iwashita T, Yilmaz OH, Terhorst C, Morrison SJ (2005) SLAM family receptors distinguish hematopoietic stem and progenitor cells and reveal endothelial niches for stem cells. Cell 121:1109–1121

Kobune F, Sakata H, Sugiura A (1990) Marmoset lymphoblastoid cells as a sensitive host for isolation of measles virus. J Virol 64:700–705

Kobune F, Takahashi H, Terao K, Ohkawa T, Ami Y, Suzaki Y, Nagata N, Sakata H, Yamanouchi K, Kai C (1996) Nonhuman primate models of measles. Lab Anim Sci 46:315–320

Kouomou DW, Wild TF (2002) Adaptation of wild-type measles virus to tissue culture. J Virol 76:1505–1509

Lawrence DM, Patterson CE, Gales TL, D'Orazio JL, Vaughn MM, Rall GF (2000) Measles virus spread between neurons requires cell contact but not CD46 expression, syncytium formation, or extracellular virus production. J Virol 74:1908–1918

Lecouturier V, Fayolle J, Caballero M, Carabana J, Celma ML, Fernandez-Munoz R, Wild TF, Buckland R (1996) Identification of two amino acids in the hemagglutinin glycoprotein of measles virus (MV) that govern hemadsorption, HeLa cell fusion, and CD46 downregulation: phenotypic markers that differentiate vaccine and wild-type MV strains. J Virol 70:4200–4204

Li L, Qi Y (2002) A novel amino acid position in hemagglutinin glycoprotein of measles virus is responsible for hemadsorption and CD46 binding. Arch Virol 147:775–786

Ma CS, Nichols KE, Tangye SG (2007) Regulation of cellular and humoral immune responses by the SLAM and SAP families of molecules. Annu Rev Immunol 25:337–379

Makhortova NR, Askovich P, Patterson CE, Gechman LA, Gerard NP, Rall GF (2007) Neurokinin-1 enables measles virus trans-synaptic spread in neurons. Virology 362:235–244

Manchester M, Valsamakis A, Kaufman R, Liszewski MK, Alvarez J, Atkinson JK, Lublin DM, Oldstone MBA (1995) Measles virus and C3 binding sites are distinct on membrane cofactor protein (CD46). Proc Natl Acad Sci U S A 92:2303–2307

Manchester M, Eto DS, Valsamakis A, Liton PB, Fernandez-Munoz R, Rota PA, Bellini WJ, Forthal DN, Oldstone MBA (2000) Clinical isolates of measles virus use CD46 as a cellular receptor. J Virol 74:3967–3974

Massé N, Ainouze M, Neel B, Wild TF, Buckland R, Langedijk JP (2004) Measles virus (MV) hemagglutinin: evidence that attachment sites for MV receptors SLAM and CD46 overlap on the globular head. J Virol 78:9051–9063

Massé N, Barrett T, Muller CP, Wild TF, Buckland R (2002) Identification of a second major site for CD46 binding in the hemagglutinin protein from a laboratory strain of measles virus (MV): potential consequences for wild-type MV infection. J Virol 76:13034–13038

Mavaddat N, Mason DW, Atkinson PD, Evans EJ, Gilbert RJ, Stuart DI, Fennelly JA, Barclay AN, Davis SJ, Brown MH (2000) Signaling lymphocytic activation molecule (CDw150) is homophilic but self-associates with very low affinity. J Biol Chem 275:28100–28109

Minagawa H, Tanaka K, Ono N, Tatsuo H, Yanagi Y (2001) Induction of the measles virus receptor (SLAM) on monocytes. J Gen Virol 82:2913–2917

Miyajima N, Takeda M, Tashiro M, Hashimoto K, Yanagi Y, Nagata K, Takeuchi K (2004) Cell tropism of wild-type measles virus is affected by amino acid substitutions in the P, V and M proteins, or by a truncation in the C protein. J Gen Virol 85:3001–3006

Moss WJ, Ryon JJ, Monze M, Cutts F, Quinn TC, Griffin DE (2002) Suppression of human immunodeficiency virus replication during acute measles. J Infect Dis 185:1035–1042

Naniche D, Wild TF, Rabourdin-Combe C, Gerlier D (1992) A monoclonal antibody recognizes a human cell surface glycoprotein involved in measles virus binding. J Gen Virol 73:2617–2624

Naniche D, Varior-Krishnan G, Cervoni F, Wild TF, Rossi B, Rabourdin-Combe C, Gerlier D (1993) Human membrane cofactor protein (CD46) acts as a cellular receptor for measles virus. J Virol 67:6025–6032

Naniche D, Yeh A, Eto D, Manchester M, Friedman RM, Oldstone MB (2000) Evasion of host defenses by measles virus: wild-type measles virus infection interferes with induction of alpha/beta interferon production. J Virol 74:7478–7484

Navaratnarajah C, Vongpunsawad S, Oezguen N, Stehle T, Braun W, Hashiguchi T, Maenaka K, Yanagi Y, Cattaneo R (2008) Dynamic interaction of the measles virus hemagglutinin with its receptor SLAM. J Biol Chem 283:11763–11771

Nielsen L, Blixenkrone-Moller M, Thylstrup M, Hansen NJ, Bolt G (2001) Adaptation of wild-type measles virus to CD46 receptor usage. Arch Virol 146:197–208

Ohno S, Seki F, Ono N, Yanagi Y (2003) Histidine at position 61 and its adjacent amino acid residues are critical for the ability of SLAM (CD150) to act as a cellular receptor for measles virus. J Gen Virol 84:2381–2388

Ohno S, Ono N, Seki F, Takeda M, Kura S, Tsuzuki T, Yanagi Y (2007) Measles virus infection of SLAM (CD150) knockin mice reproduces tropism and immunosuppression in human infection. J Virol 81:1650–1659

Ono N, Tatsuo H, Hidaka Y, Aoki T, Minagawa H, Yanagi Y (2001a) Measles viruses on throat swabs from measles patients use signaling lymphocytic activation molecule (CDw150) but not CD46 as a cellular receptor. J Virol 75:4399–4401

Ono N, Tatsuo H, Tanaka K, Minagawa H, Yanagi Y (2001b) V domain of human SLAM (CDw150) is essential for its function as a measles virus receptor. J Virol 75:1594–1600

Poste G (1971) The growth and cytopathogenicity of virulent and attenuated strains of canine distemper virus in dog and ferret macrophages. J Comp Pathol 81:49–54

Rima BK, Earle JAP, Baczko K, ter Meulen V, Liebert UG, Carstens C, Carabana J, Caballero M, Celma ML, Fernandez-Munoz R (1997) Sequence divergence of measles virus haemagglutinin during natural evolution and adaptation to cell culture. J Gen Virol 78:97–106

Santiago C, Bjorling E, Stehle T, Casasnovas JM (2002) Distinct kinetics for binding of the CD46 and SLAM receptors to overlapping sites in the measles virus hemagglutinin protein. J Biol Chem 277:32294–32301

Santibanez S, Niewiesk S, Heider A, Schneider-Schaulies J, Berbers GA, Zimmermann A, Halenius A, Wolbert A, Deitemeier I, Tischer A, Hengel H (2005) Probing neutralizing-antibody responses against emerging measles viruses (MVs): immune selection of MV by H protein-specific antibodies? J Gen Virol 86:365–374

Schneider U, von Messling V, Devaux P, Cattaneo R (2002) Efficiency of measles virus entry and dissemination through different receptors. J Virol 76:7460–7467

Schneider-Schaulies J, Schnorr J-J, Brinckmann U, Dunster L M, Baczko K, Liebert UG, Schneider-Schaulies S, ter Meulen V (1995) Receptor usage and differential downregulation of CD46 by measles virus wild-type and vaccine strains. Proc Natl Acad Sci U S A 92:3943–3947

Seki F, Ono N, Yamaguchi R, Yanagi Y (2003) Efficient isolation of wild strains of canine distemper virus in Vero cells expressing canine SLAM (CD150) and their adaptability to marmoset B95a cells. J Virol 77:9943–9950

Seki F, Takeda M, Minagawa H, Yanagi Y (2006) The recombinant wild-type measles virus containing a single N481Y substitution in its hemagglutinin cannot use a receptor CD46 as efficiently as that having the hemagglutinin of the Edmonston laboratory strain. J Gen Virol 87:1643–1648

Shibahara K, Hotta H, Katayama Y, Homma M (1994) Increased binding activity of measles virus to monkey red blood cells after long-term passage in Vero cell cultures. J Gen Virol 75:3511–3516

Shingai M, Ayata M, Ishida H, Matsunaga I, Katayama Y, Seya T, Tatsuo H, Yanagi Y, Ogura H (2003) Receptor use by vesicular stomatitis virus pseudotypes with glycoproteins of defective variants of measles virus isolated from brains of patients with subacute sclerosing panencephalitis. J Gen Virol 84:2133–2143

Sidorenko SP, Clark EA (2003) The dual-function CD150 receptor subfamily: the viral attraction. Nat Immunol 4:19–24

Tahara M, Takeda M, Yanagi Y (2005) Contributions of matrix and large protein genes of the measles virus Edmonston strain to growth in cultured cells as revealed by recombinant viruses. J Virol 79:15218–15225

Tahara M, Takeda M, Seki F, Hashiguchi T, Yanagi Y (2007a) Multiple amino acid substitutions in hemagglutinin are necessary for wild-type measles virus to acquire the ability to use receptor CD46 efficiently. J Virol 81:2564–2572

Tahara M, Takeda M, Yanagi Y (2007b) Altered interaction of the matrix protein with the cytoplasmic tail of hemagglutinin modulates measles virus growth by affecting virus assembly and cell-cell fusion. J Virol 81:6827–6836

Tahara M, Takeda M, Shirogane Y, Hashiguchi T, Ohno S, Yanagi Y (2008) Measles virus infects both polarized epithelial and immune cells using distinctive receptor-binding sites on its hemagglutinin. J Virol 82:4630–4637

Takeda M, Tahara M, Hashiguchi T, Sato TA, Jinnouchi F, Ueki S, Ohno S, Yanagi Y (2007) A human lung carcinoma cell line supports efficient measles virus growth and syncytium formation via a SLAM- and CD46-independent mechanism. J Virol 81:12091–12096

Takeuchi K, Miyajima N, Kobune F, Tashiro M (2000) Comparative nucleotide sequence analysis of the entire genomes of B95a cell-isolated and Vero cell-isolated measles viruses from the same patient. Virus Genes 20:253–257

Takeuchi K, Takeda M, Miyajima N, Kobune F, Tanabayashi K, Tashiro M (2002) Recombinant wild-type and Edmonston strain measles viruses bearing heterologous H proteins: role of H protein in cell fusion and host cell specificity. J Virol 76:4891–4900

Takeuchi K, Miyajima N, Nagata N, Takeda M, Tashiro M (2003) Wild-type measles virus induces large syncytium formation in primary human small airway epithelial cells by a SLAM (CD150)-independent mechanism. Virus Res 94:11–16

Tanaka K, Xie M, Yanagi Y (1998) The hemagglutinin of recent measles virus isolates induces cell fusion in a marmoset cell line, but not in other CD46-positive human and monkey cell lines, when expressed together with the F protein. Arch Virol 143:213–225

Taqi AM, Abdurrahman MB, Yakubu AM, Fleming AF (1981) Regression of Hodgkin's disease after measles. Lancet i:1112

Tatsuo H, Okuma K, Tanaka K, Ono N, Minagawa H, Takade A, Matsuura Y, Yanagi Y (2000a) Virus entry is a major determinant of cell tropism of Edmonston and wild-type strains of measles virus as revealed by vesicular stomatitis virus pseudotypes bearing their envelope proteins. J Virol 74:4139–4145

Tatsuo H, Ono N, Tanaka K, Yanagi Y (2000b) SLAM (CDw150) is a cellular receptor for measles virus. Nature 406:893–897

Tatsuo H, Ono N, Yanagi Y (2001) Morbilliviruses use signaling lymphocyte activation molecules (CD150) as cellular receptors. J Virol 75:5842–5850

Veillette A (2006) Immune regulation by SLAM family receptors and SAP-related adaptors. Nat Rev Immunol 6:56–66

von Messling V, Springfeld C, Devaux P, Cattaneo R (2003) A ferret model of canine distemper virus virulence and immunosuppression. J Virol 77:12579–12591

von Messling V, Milosevic D, Cattaneo R (2004) Tropism illuminated: lymphocyte-based pathways blazed by lethal morbillivirus through the host immune system. Proc Natl Acad Sci U S A 101:14216–14221

von Messling V, Svitek N, Cattaneo R (2006) Receptor (SLAM [CD150]) recognition and the V protein sustain swift lymphocyte-based invasion of mucosal tissue and lymphatic organs by a morbillivirus. J Virol 80:6084–6092

Vongpunsawad S, Oezgun N, Braun W, Cattaneo R (2004) Selectively receptor-blind measles viruses: identification of residues necessary for SLAM- or CD46-induced fusion and their localization on a new hemagglutinin structural model. J Virol 78:302–313

Wang N, Satoskar A, Faubion W, Howie D, Okamoto S, Feske S, Gullo C, Clarke K, Sosa MR, Sharpe AH, Terhorst C (2004) The cell surface receptor SLAM controls T cell and macrophage functions. J Exp Med 199:1255–1264

Woelk CH, Jin L, Holmes EC, Brown DW (2001) Immune and artificial selection in the haemagglutinin (H) glycoprotein of measles virus. J Gen Virol 82:2463–2474

Xie M-F, Tanaka K, Ono N, Minagawa H, Yanagi Y (1999) Amino acid substitutions at position 481 differently affect the ability of the measles virus hemagglutinin to induce cell fusion in monkey and marmoset cells co-expressing the fusion protein. Arch Virol 144:1689–1699

Chapter 3
Measles Virus and CD46

C. Kemper and J.P. Atkinson(✉)

Contents

CD46 Discovery and Characterization	32
Structure and Isoforms	32
Tissue Distribution	35
Functions	35
Complement Regulation	35
Fertilization	36
T Cell Regulation	37
Pathogen Receptor	37
CD46 and Measles Virus Interaction	40
MV Hemagglutinin Binding Site Within CD46	40
CD46 Binding Site Within MV Hemagglutinin	41
CD46-Mediated Mechanisms of Immunosuppression in MV Infection	42
CD46-Transgenic Mouse Models	46
Future Outlook	48
References	49

Abstract Measles virus (MV) was isolated in 1954 (Enders and Peeble 1954). It is among the most contagious of viruses and a leading cause of mortality in children in developing countries (Murray and Lopez 1997; Griffin 2001; Bryce et al. 2005). Despite intense research over decades on the biology and pathogenesis of the virus and the successful development in 1963 of an effective MV vaccine (Cutts and Markowitz 1994), cell entry receptor(s) for MV remained unidentified until 1993. Two independent studies showed that transfection of nonsusceptible rodent cells with human CD46 renders these cells permissive to infection with the Edmonston and Halle vaccine strains of measles virus (Dorig et al. 1993; Naniche et al. 1993). A key finding in these investigations was that MV binding and infection was inhibited by monoclonal and polyclonal antibodies to CD46. These reports established CD46 as a MV cell entry receptor. This chapter summarizes the role of CD46 in measles virus infection.

J.P. Atkinson
Division of Rheumatology, 660 S Euclid, Box 8045, St. Louis, MO 63110, USA, e-mail: jatkinso@im.wustl.edu

D.E. Griffin and M.B.A. Oldstone (eds.) *Measles – History and Basic Biology.*
© Springer-Verlag Berlin Heidelberg 2009

CD46 Discovery and Characterization

CD46 was discovered 1985 as a protein expressed on human peripheral blood mononuclear cells that bound C3b, the major opsonic and activation fragment of the complement system (Cole et al. 1985). It was initially termed gp45-70 because of its unusually broad, doublet, mobility pattern on SDS-PAGE. Subsequent functional analysis showed that gp45-70 functions as a complement regulator by serving as a cofactor for the plasma serine protease factor-I to cleave C3b and C4b (Seya et al. 1986; Yu et al. 1986). The protein was therefore renamed membrane cofactor protein (MCP) with a cluster of differentiation designation CD46 (Hadam 1989).

The complement system is a major player in the innate immune response, where it functions as a first-line defense against invading pathogens (Whaley et al. 1993). The complement system is activated by lectins and natural antibodies upon ligand binding and also serves as an independent immune system with sensing and effector activities (the alternative pathway). Once activated, it mediates microbial destruction by opsonizing microbes for adherence and internalization via phagocytic cells and through lysis (Whaley et al. 1993). It promotes the inflammatory process by the release of proinflammatory mediators, especially the C3a and C5a anaphylatoxins, which activate a wide range of cells involved in the host's immune response (Kohl and Bitter-Suermann 1993; Hawlisch and Kohl 2006), including endothelial and epithelial cells (Whaley et al. 1993). The complement system is often called a double-edged sword. Thus, while instrumental in fighting infections and promoting the immune response, it can cause damage to host tissues at a site of infection, in the setting of autoantibodies and immune complexes, or in acute and chronic injury states (Whaley et al. 1993; Walport 2001a, 2001b). To avoid undesirable damage to self, tight control is critical (Richards et al. 2007). Such control is achieved in part by two fluid-phase (factor H and C4b-binding protein) (Morgan and Harris 1999) and two membrane-bound (decay accelerating factor, DAF/CD55 and membrane cofactor protein, MCP/CD46) regulators. These proteins interact with C3 and/or C4 activation fragments through shared structural features (Morgan and Harris 1999). Also, the gene locus for these complement regulators is in a cluster that occupies an approximately 800-kb segment at 1q32 (de Cordoba et al. 1984; Holers et al. 1985; Reid et al. 1986; Hourcade et al. 1989). CD46 is in this regulators-of-complement-activation protein/gene cluster (Fig. 1A) (Cui et al. 1993).

Structure and Isoforms

The CD46 gene consists of 14 exons and 13 introns (Hourcade et al. 1989), spanning about 45 kb within the RCA gene cluster (Fig. 1B). It encodes a type I transmembrane protein (Liszewski and Atkinson 1992). The analysis of distinct cDNAs derived from cDNA libraries revealed one of the intriguing features of CD46: multiple isoforms arising from a single gene by alternative splicing (Fig. 1C) (Hourcade et al. 1989; Liszewski and Atkinson 1993). CD46 consists of four short consensus

3 Measles Virus and CD46

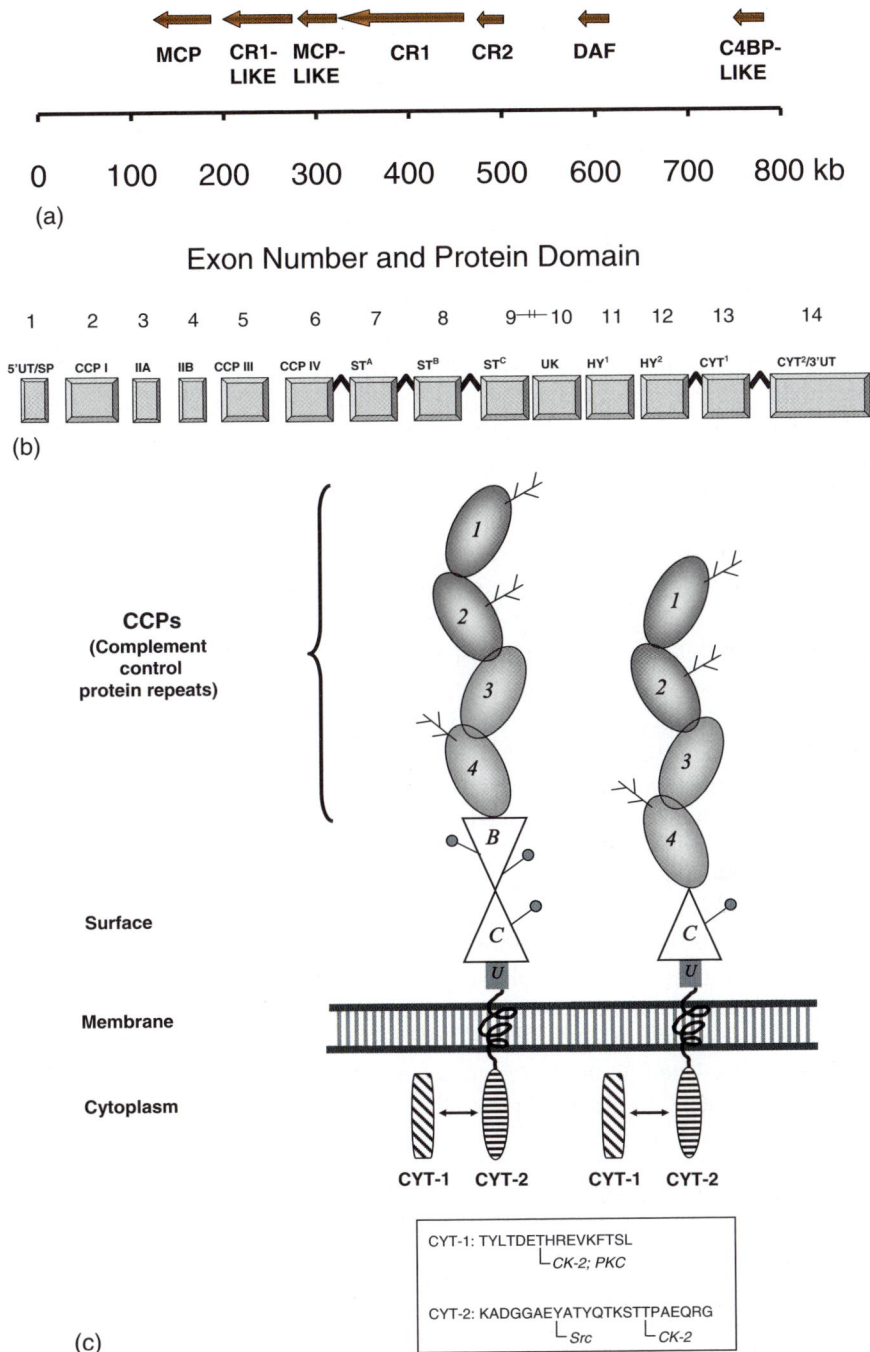

Fig. 1 (**a**) The gene sizes and intergenic distances are approximately drawn to scale. (**b**) Exon number and protein domain. The intergenic distances are not drawn to scale. The gene comprises approximately 46 kb. This is approximately 15 kb of DNA between exon 9 and 10. (**c**) Diagram of

repeats (SCRs) at its amino-terminus. These SCRs are also called complement control protein repeats (CCPs) or sushi domains and are independently folding protein modules of approximately 60 amino acids in length (Barlow et al. 1991). RCA members contain 4–30 CCPs and their C3 and C4 fragment binding sites reside within these structural units, usually requiring three CCPs to form a binding site. CCPs 1, 2, and 4 of CD46 are N-glycosylated. Glycosylation of CCP2 is essential for CD46 to function as MV receptor (Maisner and Herrler 1995; Maisner et al. 1994, 1996). The four CCPs are followed by a serine/threonine/proline (STP)-rich region. The STP region is encoded by three differentially spliced exons (giving rise to STP regions A, B, and C of 14–15 amino acids each). The STP regions are sites of O-glycosylation and the number and composition of amino acids of the STP region determines the quantity of O-glycosylation (Liszewski and Atkinson 1992). The STP region is followed by a short, juxtamembraneous 12 amino acid-long domain (separate exon) of yet unknown function, a transmembrane domain, an anchor, and one of two cytoplasmic tails, termed CYT-1 and CYT-2 (Liszewski and Atkinson 1992).

Thus, based on the observed STP splicing pattern and the distinct cytoplasmic tails, multiple CD46 isoforms can be generated (Fig. 1C). However, Northern and Western blotting and RT-PCR analyses of multiple cell lines, peripheral blood cells, and tissue samples demonstrate that CD46 is regularly expressed as variable amounts of four predominant isoforms, BC1, BC2, C1, and C2. Due to their difference in O-glycosylation, the BC1/2 isoforms show an M_r of 62,000–67,000, and the C1/2 forms have an M_r of 54,000–60,000 (Liszewski et al. 1991). The expression ratio of the four main isoforms is inherited in an autosomal codominant fashion, with three phenotypes in the population: the majority (65%) expresses predominantly the highly O-glycosylated BC1/2 forms, 6% express predominantly the less glycosylated C1/2 forms, and 29% of the population express both forms in roughly equal amounts (Liszewski et al. 1991; Wilton et al. 1992; Seya et al. 1999). All four isoforms serve as a MV receptor and binding of MV to CD46 is independent of the quantity of O-glycosylation (Maisner and Herrler 1995; Varior-Krishnan et al. 1994; Iwata et al. 1994). Soluble forms of CD46, possibly shed from the cell surface via metalloproteinases (Hakulinen and Keski-Oja 2006), are present in low concentrations in plasma, seminal fluid, and tears (Hara et al. 1992; Simpson and Holmes 1994). Their biological significance is unknown. Similarly, a role for the low frequency CD46 mRNAs encoding other isoforms, identified primarily in

---

Fig. 1 (Continued) CD46 structure. CD46 is a type I transmembrane glycoprotein that is expressed on most tissues as four major isoforms derived by alternative splicing of a single gene. The N-terminus of eachisoform consists of four complement control protein repeats CCPs, and CCPs 1, 2, and 4 each bear one N-linked complex sugar. The CCPs are followed by a serine, threonine, and proline-rich (STP) region that is O-glycosylated. The STP region, a site of alternative splicing, arises from three separate exons, designated A, B, and C. The four major isoforms of CD46 utilize the C region, whereas the B region is alternatively spliced, giving rise to either a BC or C STP region. Isoforms containing the A exon of the STP region have been reported, but are rarely observed in normal human tissue. The carboxyl terminus of CD46 is also differentially spliced, giving rise to two distinct cytoplasmic tails, designated CYT-1 (16 amino acids) and CYT-2 (23 amino acids)

EBV-transformed lymphocytes and leukemic cell lines (ABC1/2) and in the placenta (B1/2) (Hara et al. 1995; Matsumoto et al. 1992; Russell et al. 1992; Purcell et al. 1991; Johnstone et al. 1993), has not been identified.

Tissue Distribution

CD46 is expressed by nearly all nucleated cells (Seya et al. 1988). Human erythrocytes, in contrast to other primates, including the chimpanzee, do not express CD46 (Cole et al. 1985). The inherited specific CD46 expression pattern is generally identical on most all cell types in an individual (Liszewski et al. 1991). There are, however, notable exceptions. For example, in the fetal heart, CD46 is only found in the C1/2 isoforms (Gorelick et al. 1995), while the salivary gland and kidney express the -BC forms (Johnstone et al. 1993). Interestingly, sperm, kidney, salivary gland, and brain only express CYT-2 (Johnstone et al. 1993; Buchholz et al. 1996; Riley-Vargas and Atkinson 2003). Given that both tails of CD46 transduce intracellular signals upon CD46 crosslinking (see p. 42), tissue-specific expression of certain CD46 isoforms may play an important role during MV infection. Direct proof of this idea is lacking and is hampered by the fact that a suitable mouse model accurately recapitulating human MV infection is not available (see p. 46). A connection between specific CD46 isoform expression patterns and MV infection pathogenesis has not been observed. Although CD46 gene polymorphisms have been identified (Wilton et al. 1992), an analysis of a role for CD46 polymorphisms in the susceptibility to subacute sclerosing panencephalitis (SSPE) after MV infection has not shown an association (Kusuhara et al. 2000).

Functions

Complement Regulation

CD46 is an inhibitor of complement activation. It protects host cells from complement deposition by functioning as a cofactor for the factor I-mediated proteolytic inactivation of C3b and C4b (Morgan and Harris 1999; Liszewski et al. 1991) (Fig. 2). The binding sites for C3b and C4b and cofactor activity have been mapped to CCPs 2–4 (Adams et al. 1991; Iwata et al. 1995). The importance of CD46 in this process is demonstrated by the observation that individuals with CD46 haploinsufficiency secondary to mutations compromising its expression or regulatory function are predisposed to atypical or familial hemolytic uremic syndrome (HUS) (Richards et al. 2003, 2008; Kavanagh et al. 2008; Zheng and Sadler 2008). Atypical HUS is characterized by a triad of microangiopathic hemolytic anemia, thrombocytopenia, and acute renal failure (Richards et al. 2008; Caprioli et al.

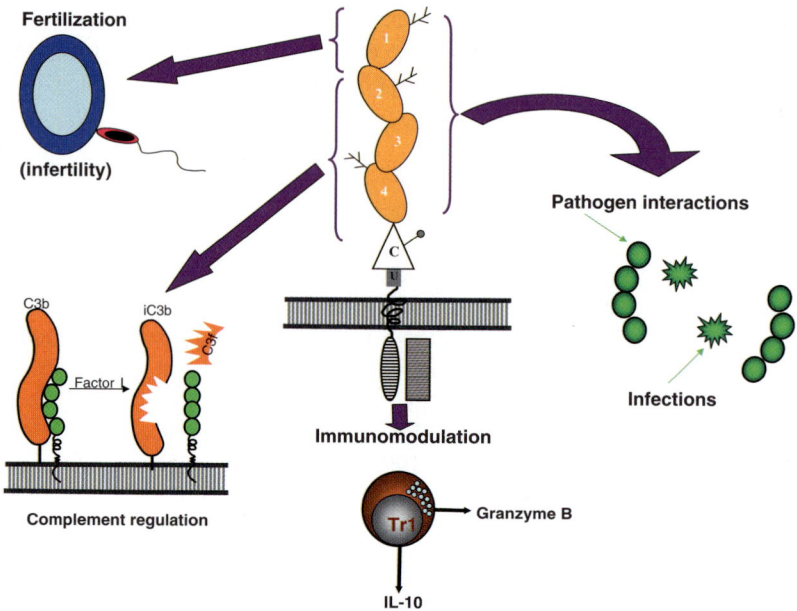

Fig. 2 Functions of CD46. CD46 was originally identified as a C3b-binding protein and then shown to have cofactor activity (promotion of the degradation of C3b and C4b by factor I) (*bottom left*). The complement regulatory activity of CD46 resides within CCPs 2–4. CD46 also plays a role in fertilization at the time sperm/egg union (*top left*). Human pathogens utilize CD46 as a cell entry receptor (*right*) (see Table 1). Concurrent activation of the T cell receptor and CD46 on primary human CD4+ T cells in the presence of IL-2 leads to the induction of IL-10 and granzyme B-producing regulatory T cells (*bottom*)

2005). A partial deficiency of CD46 is thought to lead to excessive complement activation at the site of endothelial cell injury (Caprioli et al. 2006). The resulting overexuberant innate immune and inflammatory response to injury produces a thrombomicroangiopathy of glomerular vessels.

Fertilization

CD46 serves additional functions besides complement regulation (Riley-Vargas et al. 2004). A role in male fertility was initially suggested by its expression on the inner acrosomal membrane of spermatozoa. Variations in CD46 expression levels have been associated with male infertility (Kitamura et al. 1997). CD46 is important during fertilization by presumably promoting the sperm/egg interaction (Riley-Vargas and Atkinson 2003; Riley-Vargas et al. 2004, 2005; Harris et al. 2006). Although the exact role of CD46 during fertilization is unclear (Riley-Vargas et al. 2004), another indication of its importance here is the CD46 expression pattern in rodents. In mice, rats, and guinea pigs, CD46 expression is restricted to spermatozoa (Morgan and Harris 1999; Riley-Vargas and Atkinson 2003; Tsujimura et al. 1998), while another, mouse-specific, complement regulator, Crry, takes over CD46's regulatory function on somatic tissue (Kim et al. 1995; Xu et al. 2000). The disruption of mouse CD46 causes an accelerated spontaneous acrosome reaction in sperm, suggesting that CD46 participates in this important process (Inoue et al. 2003) (Fig. 2).

T Cell Regulation

CD46 is a costimulatory molecule during T cell receptor (TCR)-mediated activation of human CD4$^+$ T lymphocytes (Astier et al. 2000; Zaffran et al. 2001). Specifically, concurrent crosslinking of CD46 with monoclonal antibodies, C3b dimers or a pathogenic ligand of CD46, along with CD3 on naïve human CD4$^+$ T cells induces a population of cells that is characterized by high IL-10 secretion (Kemper et al. 2003) and granzyme B/perforin production (Grossmann et al. 2004a, 2004b; Kemper and Atkinson 2007). CD3/CD46-activated T cells suppress the activation of bystander effector T cells through the immunosuppressive action of IL-10 and via direct killing featuring granzyme B. These characteristics place CD3/CD46-activated T cells in a regulatory T cell subpopulation. Similar to other IL-10-secreting regulatory T cells, the induction of this phenotype (including cell proliferation) (Meiffren et al. 2006) is highly dependent on the presence of exogenous IL-2 (Kemper and Atkinson 2007; Bluestone and Abbas 2003; Groux et al. 1997; Groux 2001; Roncarolo et al. 2001). Thus, the activation of T cells in the presence of complement components drives a functional phenotype distinct from that of classically CD3/CD28-activated T cells (Fig. 2).

Pathogen Receptor

CD46 is used as a receptor and port of entry by multiple human pathogens. Beside measles virus, herpes virus 6 and several adenovirus of the species B serotype utilize CD46 as cell entry receptor (Santoro et al. 1999; Cattaneo 2004; Gagar et al. 2003; Mori et al. 2002) (Fig. 2). A number of pathogenic bacteria, including *Streptococci pyogenes* as well as *Neisseria meningitides* and *N. gonorrhoeae*, bind to CD46 (Cattaneo 2004). The reason for CD46 being so attractive (pathogen magnet) for microbes is not clear yet. Obvious possibilities are that (a) CD46 could protect the invading organism from complement attack or (b) the induction of a T cell immunomodulatory phenotype. Also, in a strategy that is probably more commonly employed than currently recognized, *Escherichia coli* permit sufficient C3b deposition so as to be able to engage CD46 on epithelial cells in the urogenital system (Li et al. 2006) (Table 1).

Table 1 Pathogen/CD46 interactions and their major cellular/biological consequences

Pathogen	Ligand	Binding domain(s) within CD46	Ref	Major cellular consequences	Ref
Measles virus	Hemagglutinin (MVH)	CCPs 1–2	Manchester et al. 2000	Downregulation of IL-12 production by monocytes/macrophages	Karp et al. 1996
				Alterations in internalization pathways	Crimeen-Irwin et al. 2003
				Modulation of Th1/Th2 responses	Marie et al. 2002
				Induction of IFNα/β	Katayama et al. 2000; Kurita-Taniguchi et al. 2000
Neisseria (gonorrhoeae and meningitidis)	Type IV pilus	CCPs 3–4	Kallstrom et al. 2001	CD46 cluster formation below bacteria attachment site	Gill et al. 2003
				Ca^{2+} flux	Kallstrom et al. 1998
				Phosphorylation of CYT-2 CD46 downregulation	Gill et al. 2003; Kallstrom et al. 1998; Lee et al. 2002
Herpesvirus 6 (human)	Complex of glycoproteins H, L and Q	CCPs 2–3	Santoro et al. 2003; Mori et al. 2003; Greenstone et al. 2002	Suppression of IL-12 CD46 downregulation	Smith et al. 2003 Santoro et al. 1999
Streptococcus pyogenes	M protein	CCPs 3–4	Giannakis et al. 2002	Bacteria binding to human T cells in the presence of TCR stimulation induces IL-10 and granzyme B production (regulatory T cell development)	Price et al. 2005

Pathogen	Ligand	Binding site	References	Function	References
Adenovirus (groups B and D)	Fiberknob Ad35	CCPs 1–2	Gaggar et al. 2003; Segerman et al. 2003; Fleischli et al. 2005	Decreased pro-inflammatory cytokine production by PBMCs	Iacobelli-Martinez et al. 2005
Escherichia coli	C3b deposited on pathogenic *E. coli*	CCPs 2–4	Li et al. 2006	Unknown	
Bovine viral diarrhea virus	Unknown	CCP 1	Maurer et al. 2004; Krey et al. 2006	Virus entry	Maurer et al. 2004; Krey et al. 2006

CD46 and Measles Virus Interaction

MV Hemagglutinin Binding Site Within CD46

The MV outer envelope consists of two glycoproteins, the fusion (F) and hemagglutinin (H) protein (Griffen 2001). Both are essential for host cell entry and viral pathogenesis (Griffen 2001). While the MVF protein plays a role in cell membrane penetration and syncytium formation (Wild et al. 1994), MVH is responsible for binding of the vaccine strains Edmonston and Halle to CD46 (Maisner et al. 1994; Maisner and Herrler 1995; Devaux et al. 1996) (for a detailed description of MVF and MVH proteins see chapter 4 by C. Navaratnarajah and R. Cattaneo, this volume). MVH also binds to signaling lymphocyte-activation molecule (SOLAM, CD150), which has been identified as a second MV receptor (Tatsuo et al. 2000; Khiman et al. 2004; Erlenhofer et al. 2002) (see the chapter by Y. Yanagi et al.) (for more information on MV receptor usage see chapter 2 by Yanagi and Hashiguchi, this volume).

The MVH binding site of CD46 resides within CCPs 1 and 2 (Maisner et al. 1994; Maisner and Herrler 1995; Iwata et al. 1995; Devaux et al. 1996; Manchester et al. 1997). Although dispensable for the MV/CD46 interaction, CCP4 enhances binding of MVH to CCPs 1 and 2 (Christiansen et al. 2000). CCPs 1 and 2 each contain a single site for *N*-glycosylation. The complex sugar linked to asparagine in CCP2 is critical for MVH binding, while the carbohydrate residue in CCP1 does not influence virus–receptor interaction. Epitope mapping employing monoclonal antibodies and peptides that inhibit the MVH/CD46 interaction and functional analysis of single amino acid substitution exchanges within CCPs 1 and 2 showed that amino acids 36–59 (CCP1) and 103–118 (CCP2) of CD46 are vital in interaction of CD46 with MVH (Hsu et al. 1998; Buchholz et al. 1997). That single amino acid changes not completely abolish the CD46/MVH interaction but rather led to varying degrees of reduced binding suggests that several distinct regions in CCPs 1 and 2 cooperate in binding MVH (Casasnovas et al. 1999; Hsu et al. 1997; Lecoutrier et al. 1996; Mumenthaler et al. 1997). This notion is supported by the crystal structure and molecular modeling studies of CCPs 1 and 2 (Casasnovas et al. 1999; Mumenthaler et al. 1997).

The location of the MV-binding and the C3b-binding sites within CD46 suggest that the MV interaction may not interfere with CD46's regulation of the alternative complement pathway (Christiansen et al. 2000a, 2000b). Although the MVH recognition site partially overlaps with the C4b-binding domain of CD46 (Iwata et al. 1995), in vivo interference of MV with classical complement pathway activation has not been reported.

One specific amino acid residue of CCP1 seems to be of exceptional importance among those constituting the MVH binding site as mutation of the arginine in position 59 decreases binding of MVH to CD46 by approximately 80%. This is particularly interesting because the major families of New World monkeys lack CCP1 on somatic cells (Hsu et al. 1997). This splicing out of CCP1 may account for the low susceptibility of these species to MV infection (Hsu et al. 1997, 1998). The complement regulatory activity of monkey CD46 is not compromised as the domains for this activity reside within CCPs 2–4. In addition, the expression of CCP1 is retained in CD46 expressed on the sperm of New World monkeys (Riley et al. 2002). This manipulation of the CD46 structure by New World monkeys may represent deletion of CCP1 on peripheral blood cells to protect against infection by a major pathogen but yet conserve the vital functions of CD46 in complement regulation (CCPs 2–4) and fertilization (CCP1 in sperm). CCP1 of CD46 also harbors the binding sites for herpes virus 6 and for all CD46-binding adenoviruses. This raises the interesting question of why humans did not adapt/continue the monkey strategy and only retain CCP1 expression on sperm but delete it on somatic cells and suggests a potential yet unidentified role for CCP1 in CD46 biology besides that in sperm–egg interactions.

CD46 Binding Site Within MV Hemagglutinin

CD46 functions as a receptor for the MV vaccine strains Edmonston and Halle (Dorig et al. 1993; Naniche et al. 1993) but wild-type MV strains isolated from blood or throat swabs of patients generally do not bind to CD46 (Dhiman et al. 2004; Buckland and Wild 1997). In contrast, all known MV strains interact with SLAM (Dhiman et al. 2004; Yanagi et al. 2006; Kerdiles et al. 2006; Masse et al. 2004). Wild-type MV strains propagated in the marmoset lymphoblastic cell line B95 or Epstein-Barr virus-immortalized human B cell lines continue to only interact with SLAM (Lecouturier et al. 1996; Shibahara et al. 1994). However, wild-type strains passaged through Vero cells (Schneider-Schaulies et al. 1994) bind both receptors, CD46 and SLAM (for detailed information on MV tropisms on pathogenesis in relation to receptors, see chapter 2 by Yanagi and Hashiguchi, this volume).

Analyses of the regions within MVH responsible for this difference in receptor usage showed that only two amino acid changes at positions 546 and/or 481 determine the ability of MVH to bind to CD46 and/or SLAM. MVH (Erlenhofer et al. 2002; Hsu et al. 1998; Seki et al. 2006; Bartz et al. 1996, 1998; Rima et al. 1997; Takeuchi et al. 2002; Vongpunsawad et al. 2004; Li and Qi 2002) proteins that have a serine to glycine substitution at position 546 or a tyrosine at position 481 interact with both CD46 and SLAM. MVH containing an asparagine at the 481 position only binds SLAM. All strains analyzed after prolonged passage in Vero cells have a tyrosine at the 481 position (Yanagi et al. 2005). Because Vero cells express CD46 but not SLAM (Takeuchi et al. 2002; Johnston et al. 1999), MV is forced to adapt in Vero cell culture with the appropriate amino acid change to CD46 usage. It is, however, unclear how the wild-type virus with initial low binding and cell-entry capability overcomes this crucial entry step to propagate strongly in Vero cells.

CD46 is ubiquitously expressed in humans, suggesting that the factors that drive wild-type MV strains to favor SLAM as a receptor in vivo are likely not based on the differences in the receptor expression profiles alone. Several studies analyzing the residues within MVH relevant for the binding to CD46 identified 12 amino acids (A428, F431, V451, Y452, L464, Y481, P486, I487, A527, S546, S548, and F549) (Masse et al. 2002, 2004; Vongpunsawad et al. 2004; Santiago et al. 2002). Mapping of these amino acid residues onto the MVH crystal structure (Colf et al. 2007) (see p. 23) demonstrated that the CD46 binding site is on the rim of the so-called dead neuraminidase fold. The authors of this study also determined that the binding site for SLAM within MVH is approximately 35 Å removed from the C3b CD46 binding domain. Because of this distance, sites of these binding domains within MVH likely do not overlap (Colf et al. 2007) (for further reading on the CD46/MVH interaction, see chapter 2 by Yanagi and Hashigachi, this volume).

Interaction of MVH with a receptor induces conformational changes. These changes in MVH affect the structure of the adjacent MVF protein and trigger a series of events leading to the fusion of the viral envelope with the host cell membrane (Wild et al. 1991). Similarly, the interaction of CD46 with a pathogenic ligand also affects its structure and cell surface distribution. For example, binding of the adenovirus type 11 knob protein profoundly alters the conformation of CD46: ligand-free CD46 shows a pronounced 60-degree bend between CCPs 1 and 2. Upon binding to the adenovirus knob protein, these CCPs realign and assume a rod-like shape (Persson et al. 2007). Also, the binding of pathogenic Neisseria to CD46 on human epithelial cells induces the formation of CD46 clusters below the bacteria attachment site and CD46-dependent changes in intracellular actin distribution (Gill and Atkinson 2004). Thus, the MV–CD46 interaction triggers a complex cascade of events on both the virus and the host cell side (for a detailed description of this event see the chapter by D. Gerlier and H. Valentin, this volume). Delineating these pathways and their intricate interplay most likely holds the key to understanding the MV pathogenesis and ultimately the improvement of MV vaccines.

CD46-Mediated Mechanisms of Immunosuppression in MV Infection

The mechanisms underlying the lymphopenia and immunosuppression that accompany MV infections are not well understood (Gerlier et al. 2006; Marie et al. 2004). All MV receptors so far identified produce intracellular signals upon their engagement (Dhiman et al. 2004; Yanagi et al. 2006; Kerdiles et al. 2006; Gerlier et al. 2006). Many of these signals initiate cellular events that modulate the immune response, seemingly in favor of the pathogen (Gerlier et al. 2006; de Witte et al. 2006). A first indication that CD46 activation could be beneficial for viral dissemination was the observation that human primary monocytes downregulate IL-12p70 and p40 upon CD46 crosslinking with MVH, anti-CD46 monoclonal Abs, or C3b dimers (natural ligand) (Karp et al. 1996; Karp 1999). Since IL-12 is essential for the generation of successful effector T cell responses, MV-induced IL-12 downregulation provides an attractive mechanism for virus-mediated immunosuppression. Suppression of IL-12 synthesis has indeed been observed in MV-infected patients (Atabani et al. 2001). Early CD46-mediated signaling events in human macrophages induce recruitment of the protein-tyrosine phosphatase SHP-1 to CD46's cytoplasmic domain and then the subsequent synthesis of nitric oxide (NO) and IL-12p40 (Kurita-Taniguchi et al. 2000). The reasons for these apparently contradictory findings are unclear. One possibility is that the developmental stage of the MV-targeted cell (monocyte vs dendritic cell) elicits differential responses upon CD46 activation. Also, in one study MV was utilized as CD46 ligand while CD46 crosslinking antibodies were used in the other. Thus, signals induced upon CD46 engagement might also differ depending on the nature of the CD46-activating ligand.

MV binding to CD46 also modulates the production of another central cytokine, IFNα/β (Kurita-Taniguchi et al. 2000; Marie et al. 2001; Naniche et al. 2000). MV-exposed macrophages from huCD46-transgenic mice (see p. 46) resist infection but produce high amounts of IFNα/β (Katayama et al. 2000). The in-duction of IFNα/β is dependent on CD46-mediated signaling because macrophages expressing tail-less forms of CD46 do not produce IFNα/β and become susceptible to MV infection (Hirano et al. 2002). Little is known about the effect of the MV/CD46 interaction on B lymphocytes, but MV binding to CD46 expressed on B cells results in more efficient processing of MV antigens as well as enhanced MHC class II-restricted presentation to T cells (Gerlier et al. 1994a; Rivailler et al. 1998).

Altering the function of antigen-presenting cells such as macrophages and dendritic cells is a common strategy of many pathogens (de Witte et al. 2006). There is accumulating evidence that the MV–CD46 interaction also impacts effector T cell responses directly. The concurrent activation of CD3 and CD46 with mAbs on peripheral blood human CD4$^+$ T lymphocytes induces the production of high amounts of IL-10 and granzyme B. CD3/CD46-activated T cells acquire a phenotype reminiscent of Tr1 regulatory T cells and suppress the activation of bystander effector T cells via IL-10 and/or granzyme B (Kemper et al. 2003; Grossman et al. 2004b). T cells with these properties are predicted to aid in the contraction of an effector T cell response and in the prevention of autoimmunity (Kemper and Atkinson 2007; Bluestone and Abbas 2003; Sakaguchi 2000).

Although the in vivo role of such CD46-induced regulatory T cells is not clear, MV strains that bind CD46 may take advantage of CD46's T cell function modulatory property to perturb the protective T cell immune response and gain a foothold in the host. In accordance with this hypothesis is the observation that *Streptococci pyogenes* or the purified CD46-binding M protein of these bacteria indeed induce this suppressive Tr1 phenotype in primary human CD4$^+$ T cells (Price et al. 2005).

The notion of direct modulation of T cell responses by the MV–CD46 interaction is also supported by recent studies conducted in huCD46-transgenic mice expressing either a CYT-1 or CYT-2-bearing human CD46 isoform (see p. 46). When these animals are injected with inactivated vesicular stomatitis virus (VSV) expressing MVH, purified CD4$^+$ T cells from CYT-1-expressing animals proliferate strongly, produce IL-10 and inhibit the contact hypersensitivity reaction after concurrent TCR and CD46 activation. By contrast, CD3/CD46-activated T cells from CYT-2-expressing animals show weak proliferation, low IL-10 production, and an increased contact hypersensitivity reaction (Marie et al. 2002). The apparent difference in the signaling events induced by the two intracellular domains of CD46 suggests first that the distinct CD46 isoform pattern observed in certain tissues (see p. 32 and 35) may be of importance in MV cell entry and virus spread and second that the inherited isoform expression pattern might be associated with differences in susceptibility of a given individual to MV infection or CD46-binding pathogens in general.

A puzzling observation is that wild-type MV strains favor SLAM over CD46 as their receptor despite the ubiquitous expression profile and immunomodulatory properties of CD46 (Dhiman et al. 2004; Yanagi et al. 2006). A possible explanation is that CD4$^+$ T cells not only produce IL-10 but also the proinflammatory cytokine IFN-γ upon CD3 and CD46 engagement (Kemper et al. 2003; Sanchez et al. 2004). In addition, it is not known if CD46-induced regulatory T cells suppress both Th1 and Th2 responses. Thus, an unfavorable skewing of the effector T cell response by CD46-activated T cells might have driven MV receptor usage in vivo toward SLAM. In addition, CD46 activation by multiple ligands, including MV and MVH, on several cell types analyzed so far induces CD46 downregulation (Dhiman et al. 2004; Yanagi et al. 2006; Bartz et al. 1996; Schnoor et al. 1995; Galbraith et al. 1998). In fact, CD46 downregulation is used as a common marker for a successful or infection-propagating interaction between CD46 and MV or MVH (Schnoor et al. 1995). This is commonly viewed as a protective measure by the host, as CD46 downregulation renders MV-infected cells more vulnerable to complement-mediated lysis (Schnoor et al. 1995). Furthermore, Gasque et al. proposed that CD46 expression provides a don't-eat-me signal, much like MHC class-I (Elward et al. 2005). Loss of CD46 expression, either via CD46-activation or after induction of apoptosis, flags the cells with an eat-me signal for uptake by phagocytes (Elward et al. 2005) (for further reading on the impact of MV on host immunity, see the chapter by Gerlier and Valentin, this issue).

CD46 activation on T cells and epithelial cells induces actin skeleton rearrangements (Zaffran et al. 2001; Gill et al. 2003) and CD46 activation on T cells has been implicated in uropod formation (Oliaro et al. 2006). In addition, CYT-1 of CD46 interacts with DLG-4, a protein that provides a scaffold for signaling complexes in epithelial cells (Ludford-Menting et al. 2002). The CD46/DLG-4 interaction results in the basolateral targeting of CD46 in several different epithelial cell lines. DLG4 is also implicated in tight junction formation and membrane fusion (Ludford-Menting et al. 2002). Thus, MV binding to CD46 possibly induces changes in the actin skeleton, thereby affecting the overall cellular shape or structure of the infected cell. Since a hallmark of MV infection is the induction of cell–cell fusion with the formation of multinucleated giant cells or syncytia, it would be interesting to address if CD46 is possibly involved in this process.

The existence of an unidentified additional MV receptors has been proposed since several groups observed MV attachment and entry to cells in a SLAM and CD46-independent fashion (Hashimoto et al. 2002; Andres et al. 2003; Hall et al. 1971). The identification of this putative receptor may help to explain the characteristics of MV infection that are not well accounted for by CD46's or SLAM's expression profile and function (Table 2).

Table 2 Measles virus/CD46 interactions and cellular signaling events

Cell type	Stimulation	Major observation	Ref
Human primary peripheral blood monocytes	MV hemagglutinin, monoclonal antibodies to CD46 and C3b dimers	Downregulation of IL-12 (p70 and p40)	Karp et al. 1996
Macrophages from hCD46-transgenic mice	MV	Increase in IFN-α/β production	Kurita-Taniguchi et al. 2000; Atabani et al. 2001; Naniche et al. 2000
CD4$^+$ T cells from hCD46-transgenic mice	MV hemagglutinin	CYT-1 dependent induction of IL-10 leading to inhibition of the contact hypersensitivity reaction CYT-2 dependent increased contact hypersensitivity reaction	Marie et al. 2002
B cells from hCD46-transgenic mice	MV	Increase in MV antigen processing Enhancement of MHC class II-restricted antigen presentation	Rivailler et al. 1998; Gerlier et al. 1994

hCD46 human CD46; *CYT-1* or *-2* cytoplasmic tail of CD46

CD46-Transgenic Mouse Models

Nonhuman primates can be infected with MV experimentally and provide the animal model that most closely mimics the disease in humans (Hall et al. 1971; van Binnendijk et al. 1995). MV-infected rhesus macaques and African green monkeys show evidence of systemic viral replication, MV-induced immunosuppression, and clinical signs of disease, including maculopapular rash and conjunctivitis (Hall et al. 1971; van Binnendijk et al. 1995). On the other hand, nonhuman primate models are expensive and logistically challenging. In addition, they lack many of the advantages of small animal models, including the easy and basically unlimited access to tissue samples and the ability to study MV pathogenesis in gene knock out models. Unfortunately, the development of a successful small animal model has been hampered by the restricted function and/or expression of the two known MV receptors in rodents: mouse SLAM displays 60% structural and functional identity to human SLAM, but does not bind MV (Yanagi et al. 2006; Ono et al. 2001) and mouse CD46 is only expressed on spermatozoa (Riley-Vargas et al. 2004; Harris et al. 2006; Tsujimura et al. 1998). Similarly, rats including the cotton strain that was used in one study to analyze the function of the MV envelope outer glycoproteins in MV-induced immunosuppression (Niewiesk 1999) and guinea pigs do not express CD46 on somatic cells (Harris et al. 2006; Hosokawa et al. 1996). In addition, the intracellular domain of mouse and rat CD46 has no sequence or structure homology to either of the cytoplasmic domains of human CD46 (Hosokawa et al. 1996).

To study pathogen infections and immune reactions, several groups generated mouse strains transgenic for human CD46 (Marie et al. 2002; Mrkic et al. 1998; Rall et al. 1997; Horvat et al. 1996; Oldstone et al. 1999; Kemper et al. 2005). Some mouse strains express only one specific human CD46 isoform (Marie et al. 2002) and should be useful in analyzing functional differences between the distinct isoforms. A mouse line generated utilizing a yeast artificial chromosome that contains the complete CD46 gene, including the regulatory regions, mimics the CD46 expression pattern found in humans, including several tissue-specific distribution patterns of the four isoforms (Hourcade et al. 1990, 1992; Kemper et al. 2001). These mice have the obvious advantage that possible cooperative signaling events induced by both intracellular domains of CD46 can still occur, but it is unclear if the signaling platform and proper intermediates are present in rodents.

Intracerebral MV inoculation of mice expressing human CD46 isoforms under the control of a neuron-specific promotor induced disease and mortality, while nontransgenic control animal did not show signs of infection (Rall et al. 1997). In addition, injection of MV into the brain of these animals induced MV-associated cellular immune responses, including migration and infiltration of CD4[+] and CD8[+] T lymphocytes (Dorig et al. 1993; Rall et al. 1997; Manchester et al. 2000a). However, huCD46-transgenic mice are not susceptible to MV infection when the virus is administered via another route (for example, the natural respiratory route) than cerebral injection (Manchester and Rall 2001) because MV replication is limited (Oldstone et al. 1999; Horvat et al. 1996; Niewiesk et al. 1997; Thorley et al.

Table 3 Current hCD46-transgenic mouse strains

CD46 isoforms	Mouse model	Reference
C1	MV	Horvat et al. 1996
BC1	MV	Rall et al. 1997; Oldstone et al. 1999
C1	MV	Evlashev et al. 2000, 2001
C1 strain and a C2 strain	CD4 T cell activation/regulation	Marie et al. 2002
C1, C2, BC1, BC2 (YAC)	Xenotransplantation; MV	Oldstone et al. 1999; Yannoutsos et al. 1996
Minigene, isoforms*	Xenotransplantation; MV	Thorley et al. 1997
C1, C2, BC1, BC2 (YAC)	MV; *Neisseria meningitidis*	Mrkic et al. 1998; Johansson et al. 2003
80kb Genomic fragment, isoforms*	MV	Blixenkrone-Moeller et al. 1998
C1, C2, BC1, BC2 (YAC)	Expression study; xenotransplantation; MV	Mrkic et al. 1998; Kemper et al. 2001; Yannoutsos et al. 1996
C1 or C2 (crossed with CD 150$^{-/-}$ and IFNAR1$^{-/-}$)	Measles virus infection	Shingai et al. 2005

*Isoform expression pattern not depicted in study. YAC, yeast artificial chromosome.

1997) by the host's antiviral response. Crossing huCD46-trangenice mice with mice deficient in the type I IFN receptor (IFNAR1) (Mrkic et al. 1998) or the transcription factor STAT1 (Shingai et al. 2005) improved this model because respiratory inoculation of the virus resulted in enhanced virus spread and lung tissue inflammation (Oldstone et al. 1999). However, even in these animals, late-stage virus replication is inefficient and virus spread and MV pathogenesis only partially mimics the human disease (Peng et al. 2003).

The more limited expression of the protein in mice transgenic for human SLAM is thought to be one reason for the failure to obtain systemic MV infection in huSLAM-transgenic animals (Hahm et al. 2003, 2004). huSLAM/huCD46 double-transgenic mice, either with or without IFNAR1 expression, have been generated recently (Shingai et al. 2005). MV infection of huSLAM/huCD46 double-transgenic mice with a functional IFNAR did not induce systemic infection or disease. However, systemic infection occurred in huSLAM/huCD46/IFNAR1$^{-/-}$ animals or in huSLAM/huCD46/IFNAR1$^{+/+}$ mice injected with MV-infected DCs (Shingai et al. 2005). Thus, this study establishes that DCs and likely IFN production by these cells play a critical role in MV infection in mouse models. This is in agreement with the previous finding that MV-induced activation of CD46 expressed on human macrophages alters their cytokine profile (Karp et al. 1996) (see p. 42). It will be interesting to now delineate the roles of CD46 and SLAM in this process, specifically in the interplay between T lymphocytes and DCs.

In 2002, a study suggested an interaction between human toll-like receptor (TLR)2 and MV (Bieback et al. 2002). Wild-type MVH activates TLR2 on macrophages and monocytic cells, resulting in the production of IL-6 (Bieback et al. 2002). Given the essential role of TLRs in the recognition of pathogens (Medzhitov et al. 1997), MV infection likely triggers danger signals within this protein family. Thus, TLR knock-out mice should be considered in the generation of animal models studying MV pathogenesis.

Taken together, a number of mouse models are available to study certain aspects of MV pathogenesis. A model faithfully recapitulating the human disease does not exist. We anticipate the discovery of at least one additional human MV receptor. It does seem though that the mouse went through much effort to rid itself of MV cell entry receptors on somatic tissue. Thus, other murine factors important for viral RNA synthesis and virus assembly may also not support MV infection (Table 3).

Future Outlook

Successful immune defense is increasingly visualized as a process in which the innate part plays vital roles in instructing and guiding the adaptive system. That so many important human pathogens utilize CD46 as receptor attests to its central role at this important interface and communication between innate and adaptive immunity. Although we have learned much about CD46's functions as a measles virus receptor, its role during MV infections remains largely enigmatic. Thus, delineating the cellular mechanisms that drive the diverse CD46–pathogen interactions in the context of the immune response model systems will be key to improving our understanding of MV pathogenesis.

References

Adams EM, Brown MC, Nunge M, Krych M, Atkinson JP (1991) Contribution of the repeating domains of membrane cofactor protein (CD46) of the complement system to ligand binding and cofactor activity. J Immunol 147:3005–3011
Andres O, Obojes K, Kim KS, ter Meulen V, Schneider-Schaulies J (2003) CD46- and CD 150-independent endothelial cell infection with wild-type measles viruses. J Gen Virol 84:1189–1197
Astier A, Trescol-Biemont M-C, Azocar O, Lamouille B, Rabourdin-Combe C (2000) Cutting edge: CD46, a new costimulatory molecule for T cells, that induces p120^{CBL} and LAT phosphorylation. J Immunol 164:6091–6095
Atabani SF, Byrnes AA, Jaye A et al (2001) Natural measles causes prolonged suppression of interleukin-12 production. J Infect Dis 184:1–9
Barlow PN, Baron M, Norman DG et al (1991) Secondary structure of a complement control protein module by two-dimensional 1H NMR. Biochemistry (Mosc) 30:997–1004
Bartz R, Brinckmann U, Dunster LM, Rima B, Ter Meulen V, Schneider-Schaulies J (1996) Mapping amino acids of the measles virus hemagglutinin responsible for receptor (CD46) downregulation. Virology 224:334–337
Bartz R, Firsching R, Rima B, ter Meulen V, Schneider-Schaulies J (1998) Differential receptor usage by measles virus strains. J Gen Virol 79:1015–1025
Bieback K, Lien E, Klagge IM et al (2002) Hemagglutinin protein of wild-type measles virus activates toll-like receptor 2 signaling. J Virol 76:8729–8736
Blixenkrone-Moeller M, Bernard A, Bencsik A et al (1998) Role of CD46 in measles virus infection in CD46 transgenic mice. Virology 249:238–248
Bluestone JA, Abbas AK (2003) Natural versus adaptive regulatory T cells. Nat Rev Immunol 3:253–257
Bryce J, Boschi-Pinto C, Shibuya K, Black RE (2005) WHO estimates of the causes of death in children. Lancet 365(9465):1147–1152
Buchholz CJ, Gerlier D, Hu A et al (1996) Selective expression of a subset of measles virus receptor-competent CD46 isoforms in human brain. Virology 217:349–355
Buchholz CJ, Koller D, Devaux P et al (1997) Mapping of the primary binding site of measles virus to its receptor CD46. J Biol Chem 272:22072–22079
Buckland R, Wild TF (1997) Is CD46 the cellular receptor for measles virus? Virus Res 48:1–9
Caprioli J, Peng L, Remuzzi G (2005) The hemolytic uremic syndromes. Curr Opin Crit Care 11:487–492
Caprioli J, Noris M, Brioschi S et al (2006) Genetics of HUS: the impact of MCP, CFH and IF mutations on clinical presentation, response to treatment, and outcome. Blood 108: 1267–1279
Casasnovas JM, Larvie M, Stehle T (1999) Crystal structure of two CD46 domains reveals an extended measles virus-binding surface. EMBO J 18:2911–2922
Cattaneo R (2004) Four viruses, two bacteria, and one receptor: membrane cofactor protein (CD46) as pathogens' magnet. J Virol 78:4385–4388
Christiansen D, Deleage G, Gerlier D (2000a) Evidence for distinct complement regulatory and measles virus binding sites on CD46 SCR2. Eur J Immunol 30:3457–3462

Christiansen D, Devaux P, Reveil B et al (2000b) Octamerization enables soluble CD46 receptor to neutralize measles virus in vitro and in vivo. J Virol 74:4672–4678

Cole JL, Housley GA Jr, Dykman TR, MacDermott RP, Atkinson JP (1985) Identification of an additional class of C3-binding membrane proteins of human peripheral blood leukocytes and cell lines. Proc Natl Acad Sci U S A 82:859–863

Colf LA, Juo ZS, Garcia KC (2007) Structure of the measles virus hemagglutinin. Nat Struct Mol Biol 14:1227–1228

Crimeen-Irwin B, Ellis S, Christiansen D et al (2003) Ligand binding determines whether CD46 is internalized by clathrin-coated pits or macropinocytosis. J Biol Chem 278:46927–46937

Cui W, Hourcade D, Post TW, Greenlund AC, Atkinson JP, Kumar V (1993) Characterization of the promoter region of the membrane cofactor protein (CD46) gene of the human complement system and comparison to a membrane cofactor protein-like genetic element. J Immunol 151:4137–4146

Cutts FT, Markowitz LE (1994) Successes and failures in measles control. J Infect Dis 170 [Suppl 1]:S32–S41

de Cordoba SR, Dykman TR, Ginsberg-Fellner F et al (1984) Evidence for linkage between the loci coding for the binding protein for the fourth component of human complement (C4BP) and for the C3b/C4b receptor. Proc Natl Acad Sci U S A 81:7890–7892

de Witte L, Abt M, Schneider-Schaulies S, van Kooyk Y, Geijtenbeek TB (2006) Measles virus targets DC-SIGN to enhance dendritic cell infection. J Virol 80:3477–3486

Devaux P, Loveland BE, Christiansen D, Millane J, Gerlier D (1996) Interaction between the ectodomains of haemagglutinin and CD46 as a primary step in measles virus entry. J Gen Virol 77:1477–1481

Dhiman N, Jacobson RM, Poland GA (2004) Measles virus receptors: SLAM and CD46. Rev Med Virol 14:217–229

Dorig RE, Marcil A, Chopra A, Richardson CD (1993) The human CD46 molecule is a receptor for measles virus (Edmonston strain). Cell 75:295–305

Elward K, Griffiths M, Mizuno M et al (2005) CD46 plays a key role in tailoring innate immune recognition of apoptotic and necrotic cells. J Biol Chem 280:36342–3654

Enders JF, Peebles TC (1954) Propagation in tissue cultures of cytopathogenic agents from patients with measles. Proc Soc Exp Biol Med 86:277–286

Erlenhofer C, Duprex WP, Rima BK, ter Meulen V, Schneider-Schaulies J (2002) Analysis of receptor (CD46, CD150) usage by measles virus. J Gen Virol 83:1431–1436

Evlashev A, Moyse E, Valentin H et al (2000) Productive measles virus brain infection and apoptosis in CD46 transgenic mice. J Virol 74:1373–1382

Evlashev A, Valentin H, Rivailler P, Azocar O, Rabourdin-Combe C, Horvat B (2001) Differential permissivity to measles virus infection of human and CD46-transgenic murine lymphocytes. J Gen Virol 82:2125–2129

Fleischli C, Verhaagh S, Havenga M, et al (2005) The distal short consensus repeats 1 and 2 of the membrane cofactor protein CD46 and their distance from the cell membrane determine productive entry of species B adenovirus serotype 35. J Virol 79:10013–10022

Gaggar A, Shayakhmetov DM, Lieber A (2003) CD46 is a cellular receptor for group B adenoviruses. Nat Med 9:1408–1412

Galbraith SE, Tiwari A, Baron MD, Lund BT, Barrett T, Cosby SL (1998) Morbillivirus downregulation of CD46. J Virol 72:10292–10297

Gerlier D, Loveland BE, Varion-Krishnan G, Thorley B, McKenzie IFC, Rabaudin-Combe C (1994a) Efficient major histocompatibility complex class II-restricted presentation of measles virus relies on hemagglutinin-mediated targeting to its cellular receptor human CD46 expressed by murine B cell. J Exp Med 179:353–358

Gerlier D, Trescol-Biemont MC, Varior-Krishnan G, Naniche D, Fugier-Vivier I, Rabourdin-Combe C (1994b) Efficient major histocompatibility complex class II-restricted presentation of measles virus relies on hemagglutinin-mediated targeting to its cellular receptor human CD46 expressed by murine B cells. J Exp Med 179:353–358

Gerlier D, Valentin H, Laine D, Rabourdin-Combe C, Servet-Delprat C (2006) Subversion of the immune system by measles virus: a model for the intricate interplay between a virus and the human immune system. In: Lachmann PJ, Oldstone MBA (eds) Microbial subversion of immunity: current topics. Caister Academic, Norfolk, UK, pp 225–292

Giannakis E, Jokiranta TS, Ormsby RJ et al (2002) Identification of the streptococcal M protein binding site on membrane cofactor protein (CD46). J Immunol 168:4585–4592

Gill D, Atkinson J (2004) CD46 in *Neisseria pathogenesis*. Trends Mol Med 10:459–465

Gill DB, Koomey M, Cannon JG, Atkinson JP (2003) Down-regulation of CD46 by piliated *Neisseria gonorrhoeae*. J Exp Med 198:1313–1322

Gorelick A, Oglesby TJ, Rashbaum W, Atkinson JP, Buyon JP (1995) Ontogeny of membrane cofactor protein: phenotypic divergence in the fetal heart. Lupus 4:293–296

Greenstone HL, Santoro F, Lusso P, Berger EA (2002) Human herpesvirus 6 and measles virus employ distinct CD46 domains for receptor function. J Biol Chem 277:39112–39118

Griffin DE (2001) Measles virus. In: Knipe DM, Howley PM, Griffin DE et al (eds) Fields virology. 4th edn. Lippincott Williams & Wilkins, Philadelphia, pp 1401–1441

Grossman WJ, Verbsky JW, Barchet W, Colonna M, Atkinson JP, Ley TJ (2004a) Human T regulatory cells can use the perforin pathway to cause autologous target cell death. Immunity 21:589–601

Grossman WJ, Verbsky JW, Tollefsen BL, Kemper C, Atkinson JP, Ley TJ (2004b) Differential expression of granzymes A and B in human cytotoxic lymphocyte subsets and T regulatory cells. Blood 104:2840–2848

Groux H (2001) An overview of regulatory T cells. Microbes Infect 3:883–889

Groux H, O'Garra A, Bigler M et al (1997) A CD4$^+$ T-cell subset inhibits antigen-specific T-cell responses and prevents colitis. Nature 389:737–742

Hadam MR (1989) Cluster report: CD. In: Knapp W (ed) Leucocyte type IV. Oxford Universtiy Press, Oxford, pp 649–652

Hahm B, Arbour N, Naniche D, Homann D, Manchester M, Oldstone MB (2003) Measles virus infects and suppresses proliferation of T lymphocytes from transgenic mice bearing human signaling lymphocytic activation molecule. J Virol 77:3505–3515

Hahm B, Arbour N, Oldstone MB (2004) Measles virus interacts with human SLAM receptor on dendritic cells to cause immunosuppression. Virology 323:292–302

Hakulinen J, Keski-Oja J (2006) ADAM10-mediated release of complement membrane cofactor protein during apoptosis of epithelial cells. J Biol Chem 281:21369–1376

Hall WC, Kovatch RM, Herman PH, Fox JG (1971) Pathology of measles in rhesus monkeys. Vet Pathol 8:307–319

Hara T, Kuriyama S, Kiyohara H, Nagase Y, Matsumoto M, Seya T (1992) Soluble forms of membrane cofactor protein (CD46, MCP) are present in plasma, tears, and seminal fluid in normal subjects. Clin Exp Immunol 89:490–494

Hara T, Suzuki Y, Semba T, Hatanaka M, Matsumoto M, Seya T (1995) High expression of membrane cofactor protein of complement (CD46) in human leukaemia cell lines: implication of an alternatively spliced form containing the STA domain in CD46 up-regulation. Scand J Immunol 42:581–590

Harris CL, Mizuno M, Morgan BP (2006) Complement and complement regulators in the male reproductive system. Mol Immunol 43:57–67

Hashimoto K, Ono N, Tatsuo H et al (2002) SLAM (CD150)-independent measles virus entry as revealed by recombinant virus expressing green fluorescent protein. J Virol 76:6743–6749

Hawlisch H, Kohl J (2006) Complement and Toll-like receptors: key regulators of adaptive immune responses. Mol Immunol 43:13–21

Hirano A, Kurita-Taniguchi M, Katayama Y, Matsumoto M, Wong TC, Seya T (2002) Ligation of human CD46 with purified complement C3b or F(ab')$_2$ of monoclonal antibodies enhances isoform-specific interferon gamma-dependent nitric oxide production in macrophages. J Biochem (Tokyo) 132:83–91

Holers VM, Cole JL, Lublin DM, Seya T, Atkinson JP (1985) Human C3b- and C4b-regulatory proteins: a new multi-gene family. Immunol Today 6:188–192

Horvat B, Rivailler P, Varior-Krishnan G, Cardoso A, Gerlier D, Rabourdin-Combe C (1996) Transgenic mice expressing human measles virus (MV) receptor CD46 provide cells exhibiting different permissivities to MV infections. J Virol 70:6673–6681

Hosokawa M, Nonaka M, Okada N, Okada H (1996) Molecular cloning of guinea pig membrane cofactor protein: preferential expression in testis. J Immunol 157:4946–4952

Hourcade D, Holers VM, Atkinson JP (1989) The regulators of complement activation (RCA) gene cluster. Adv Immunol 45:381–416

Hourcade D, Liszewski MK, Kruch-Goldberg M, Atkinson SP (2000) Functional domains, structural variations and pathogen interactions of MCP, DAF, and CRI. Immunopharmacology 49:103–116

Hourcade D, Post TW, Holers VM, Lublin DM, Atkinson JP (1990) Polymorphisms of the regulators of complement activation (RCA) gene cluster. Complement Inflamm 17:302–314

Hsu EC, Dorig RE, Sarangi F, Marcil A, Iorio C, Richardson CD (1997) Artificial mutations and natural variations in the CD46 molecules from human and monkey cells define regions important for measles virus binding. J Virol 71:6144–6154

Hsu EC, Sarangi F, Iorio C et al (1998) A single amino acid change in the hemagglutinin protein of measles virus determines its ability to bind CD46 and reveals another receptor on marmoset B cells. J Virol 72:2905–2916

Iacobelli-Martinez M, Nepomuceno RR, Connolly J, Nemerow GR (2005) CD46-utilizing adenoviruses inhibit C/EBPbeta-dependent expression of proinflammatory cytokines. J Virol 79:11259–11268

Inoue N, Ikawa M, Nakanishi T et al (2003) Disruption of mouse CD46 causes an accelerated spontaneous acrosome reaction in sperm. Mol Cell Biol 23:2614–2622

Iwata K, Seya T, Ueda S, Ariga H, Nagasawa S (1994) Modulation of complement regulatory function and measles virus receptor function by the serine-threonine-rich domains of membrane cofactor protein (CD46). Biochem J 304:169–175

Iwata K, Seya T, Yanagi Y et al (1995) Diversity of sites for measles virus binding and for inactivation of complement C3b and C4b on membrane cofactor protein CD46. J Biol Chem 270:15148–15152

Johansson L, Rytkonen A, Bergman P et al (2003) CD46 in meningococcal disease. Science 301:373–375

Johnston IC, ter Meulen V, Schneider-Schaulies J, Schneider-Schaulies S (1999) A recombinant measles vaccine virus expressing wild-type glycoproteins: consequences for viral spread and cell tropism. J Virol 73:6903–6915

Johnstone RW, Russell SM, Loveland BE, McKenzie IF (1993) Polymorphic expression of CD46 protein isoforms due to tissue-specific RNA splicing. Mol Immunol 30:1231–1241

Kallstrom H, Islam MS, Berggren PO, Jonsson AB (1998) Cell signaling by the type IV pili of pathogenic *Neisseria*. J Biol Chem 273:21777–21782

Kallstrom H, Blackmer Gill D, Albiger B, Liszewski MK, Atkinson JP, Jonsson AB (2001) Attachment of *Neisseria gonorrhoeae* to the cellular pilus receptor CD46: identification of domains important for bacterial adherence. Cell Microbiol 3:133–143

Karp CL (1999) Measles: immunosuppression, interleukin-12, and complement receptors. Immunol Rev 168:91–101

Karp CL, Wysocka M, Wahl LM et al (1996) Mechanism of suppression of cell-mediated immunity by measles virus. [Erratum appears in Science 1997 Feb 21; 275(5303):1053]. Science 273:228–231

Katayama Y, Hirano A, Wong TC (2000) Human receptor for measles virus (CD46) enhances nitric oxide production and restricts virus replication in mouse macrophages by modulating production of alpha/beta interferon. J Virol 74:1252–1257

Kavanagh D, Richards A, Atkinson J (2008) Complement regulatory genes and hemolytic uremic syndromes. Annu Rev Med 59:293–309

Kemper C, Atkinson JP (2007) T-cell regulation: with complements from innate immunity. Nat Rev Immunol 7:9–18

Kemper C, Leung M, Stephensen CB et al (2001) Membrane cofactor protein (MCP; CD46) expression in transgenic mice. Clin Exp Immunol 124:180–189

Kemper C, Chan AC, Green JM, Brett KA, Murphy KM, Atkinson JP (2003) Activation of human CD4+ cells with CD3 and CD46 induces a T-regulatory cell 1 phenotype. Nature 421:388–392

Kemper C, Verbsky JW, Price JD, Atkinson JP (2005) T-cell stimulation and regulation: with complements from CD46. Immunol Res 32:31–43

Kerdiles YM, Sellin CI, Druelle J, Horvat B (2006) Immunosuppression caused by measles virus: role of viral proteins. Rev Med Virol 16:49–63

Kim YU, Kinoshita T, Molina H et al (1995) Mouse complement regulatory protein Crry/p65 uses the specific mechanisms of both human decay accelerating factor and membrane cofactor protein. J Exp Med 181:151–159

Kitamura M, Matsumiya K, Yamanaka M et al (1997) Possible association of infertility with sperm-specific abnormality of CD46. J Reprod Immunol 33:83–88

Kohl J, Bitter-Suermann D (1993) Anaphylatoxins. In: Whaley K, Loos M, Weiler JM, (eds) Immunology and medicine: complement in health disease. 2nd edn. Kluwer Academic, Boston, pp 299–324

Krey T, Himmelreich A, Heimann M et al (2006) Function of bovine CD46 as a cellular receptor for bovine viral diarrhea virus is determined by complement control protein 1. J Virol 80:3912–3922

Kurita-Taniguchi M, Fukui A, Hazeki K et al (2000) Functional modulation of human macrophages through CD46 (measles virus receptor): production of IL-12 p40 and nitric oxide in association with recruitment of protein-thyroxine phosphatase SHP-1 to CD46. J Immunol 165:5143–5152

Kusuhara K, Sasaki Y, Nakao F et al (2000) Analysis of measles virus binding sites of the CD46 gene in patients with subacute sclerosing panencephalitis. J Infect Dis 181:1447–1449

Lecouturier V, Fayolle J, Caballero M et al (1996) Identification of two amino acids in the hemagglutinin glycoprotein of measles virus (MV) that govern hemadsorption, HeLa cell fusion, and CD46 downregulation: phenotypic markers that differentiate vaccine and wild-type MV strains. J Virol 70:4200–4204

Lee SW, Bonnah RA, Higashi DL, Atkinson JP, Milgram SL, So M (2002) CD46 is phosphorylated at tyrosine 354 upon infection of epithelial cells by *Neisseria gonorrhoeae*. J Cell Biol 156:951–957

Li K, Feito MJ, Sacks SH, Sheerin NS (2006) CD46 (membrane cofactor protein) acts as a human epithelial cell receptor for internalization of opsonized uropathogenic *Escherichia coli*. J Immunol 177:2543–2551

Li L, Qi Y (2002) A novel amino acid position in hemagglutinin glycoprotein of measles virus is responsible for hemadsorption and CD46 binding. Arch Virol 147:775–786

Liszewski MK, Atkinson JP (1992) Membrane cofactor protein. Curr Top Microbiol Immunol 178:45–60

Liszewski MK, Farries TC, Lublin DM, Rooney IA, Atkinson JP (1996) Control of the complement system. Adv Immunol 41:201–283

Liszewski MK, Post TW, Atkinson JP (1991) Membrane cofactor protein (MCP or CD46): newest member of the regulators of complement activation gene cluster. Annu Rev Immunol 9:431–455

Ludford-Menting MJ, Thomas SJ, Crimeen B et al (2002) A functional interaction between CD46 and DLG4: a role for DLG4 in epithelial polarization. J Biol Chem 277:4477–4484

Maisner A, Herrler G (1995) Membrane cofactor protein with different types of N-glycans can serve as measles virus receptor. Virology 210:479–481

Maisner A, Schneider-Schaulies J, Liszewski MK, Atkinson JP, Herrler G (1994) Binding of measles virus to membrane cofactor protein (CD46): importance of disulfide bonds and N-glycans for the receptor function. J Virol 68:6299–6304

Maisner A, Alvarez J, Liszewski MK, Atkinson JP, Atkinson D, Herrler G (1996) The N-glycan of the SCR 2 region is essential for membrane cofactor protein (CD46) to function as measles virus receptor. J Virol 70:4973–4977

Manchester M, Gairin JE, Alvarez J, Liszewski MK, Atkinson JP, Oldstone MBA (1997) Measles virus recognizes its receptor, CD46, via two distinct binding domains within SCR1–2. Virology 233:174–187

Manchester M, Rall GF (2001) Model systems: transgenic mouse models for measles pathogenesis. Trends Microbiol 9:19–23

Manchester M, Eto DS, Valsamakis A et al (2000a) Clinical isolates of measles virus use CD46 as a cellular receptor. J Virol 74:3967–3974

Manchester M, Naniche D, Stehle T (2000b) CD46 as a measles receptor: form follows function. Virology 274:5–10

Marie JC, Kehren J, Trescol-Biemont MC et al (2001) Mechanism of measles virus-induced suppression of inflammatory immune responses. Immunity 14:69–79

Marie J, Astier AL, Rivailler P, Rabourdin-Combe C, Wild TF, Horvat B (2002) Linking innate and acquired immunity: divergent role of CD46 cytoplasmic domains in T cell-induced inflammation. Nat Immunol 3:659–666

Masse N, Barrett T, Muller CP, Wild TF, Buckland R (2002) Identification of a second major site for CD46 binding in the hemagglutinin protein from a laboratory strain of measles virus (MV): potential consequences for wild-type MV infection. J Virol 76:13034–13038

Masse N, Ainouze M, Neel B, Wild TF, Buckland R, Langedijk JP (2004) Measles virus (MV) hemagglutinin: evidence that attachment sites for MV receptors SLAM and CD46 overlap on the globular head. J Virol 78:9051–9063

Matsumoto M, Seya T, Nagasawa S (1992) Polymorphism and proteolytic fragments of granulocyte membrane cofactor protein (MCP, CD46) of complement. Biochem J 281:493–499

Maurer K, Krey T, Moennig V, Thiel HJ, Rumenapf T (2004) CD46 is a cellular receptor for bovine viral diarrhea virus. J Virol 78:1792–1799

Medzhitov R, Preston-Hurlburt P, Janeway CAJ (1997) A human homologue of the *Drosophilia* Toll protein signals activation of adaptive immunity. Nature 388:394–397

Meiffren G, Flacher M, Azocar O, Rabourdin-Combe C, Faure M (2006) Cutting edge: abortive proliferation of CD46-induced Tr1-like cells due to a defective Akt/Survivin signaling pathway. J Immunol 177:4957–4961

Morgan BP, Harris CL (1999) Complement regulatory proteins. Academic Press, New York

Mori Y, Seya T, Huang HL, Akkapaiboon P, Dhepakson P, Yamanishi K (2002) Human herpesvirus 6 variant A but not variant B induces fusion from without in a variety of human cells through a human herpesvirus 6 entry receptor, CD46. J Virol 76:6750–6761

Mori Y, Yang X, Akkapaiboon P, Okuno T, Yamanishi K (2003) Human herpesvirus 6 variant A glycoprotein H-glycoprotein L-glycoprotein Q complex associates with human CD46. J Virol 77:4992–4999

Mrkic B, Pavlovic J, Rulicke T et al (1998) Measles virus spread and pathogenesis in genetically modified mice. J Virol 72:7420–7427

Mumenthaler C, Schneider U, Buchholz CJ, Koller D, Braun W, Cattaneo R (1997) A 3D model for the measles virus receptor CD46 based on homology modeling, Monte Carlo simulations, and hemagglutinin binding studies. Protein Sci 6:588–597

Murray CJ, Lopez AD (1997) Alternative projections of mortality and disability by cause 1990–2020: Global Burden of Disease Study. Lancet 349(9064):1498–1504

Naniche D, Varior-Krishnan G, Cervoni F et al (1993) Human membrane cofactor protein (CD46) acts as a cellular receptor for measles virus. J Virol 67:6025–6032

Naniche D, Yeh A, Eto D, Manchester M, Friedman RM, Oldstone MB (2000) Evasion of host defenses by measles virus: wild-type measles virus infection interferes with induction of alpha/beta interferon production. J Virol 74:7478–7484

Niewiesk S (1999) Cotton rats (*Sigmodon hispidus*): an animal model to study the pathogenesis of measles virus infection. Immunol Lett 65:47–50

Niewiesk S, Schneider-Schaulies J, Ohnimus H et al (1997) CD46 expression does not overcome the intracellular block of measles virus replication in transgenic mice. J Virol 71:7969–7973

Oldstone MB, Lewicki H, Thomas D et al (1999) Measles virus infection in a transgenic model: virus-induced immunosuppression and central nervous system disease. Cell 98:629–640

Oliaro J, Pasam A, Waterhouse NJ et al (2006) Ligation of the cell surface receptor, CD46, alters T cell polarity and response to antigen presentation. Proc Natl Acad Sci U S A 103:18685–18690

Ono N, Tatsuo H, Tanaka K, Minagawa H, Yanagi Y (2001) V domain of human SLAM (CDw150) is essential for its function as a measles virus receptor. J Virol 75:1594–1600

Peng KW, Frenzke M, Myers R et al (2003) Biodistribution of oncolytic measles virus after intraperitoneal administration into Ifnar-CD46Ge transgenic mice. Hum Gene Ther 14:1565–1577

Persson BD, Reiter DM, Marttila M et al (2007) Adenovirus type 11 binding alters the conformation of its receptor CD46. Nat Struct Mol Biol 14:164–166

Price JD, Schaumburg J, Sandin C, Atkinson JP, Lindahl G, Kemper C (2005) Induction of a regulatory phenotype in human CD4$^+$ T cells by streptococcal M protein. J Immunol 175:677–684

Purcell DFJ, Russell SM, Deacon NJ, Brown MA, Hooker DJ, McKenzie IFC (1991) Alternatively spliced RNAs encode several isoforms of CD46 (MCP), a regulator of complement activation. Immunogenetics 33:335–344

Rall GF, Manchester M, Daniels LR, Callahan EM, Belman AR, Oldstone MB (1997) A transgenic mouse model for measles virus infection of the brain. Proc Natl Acad Sci U S A 94:4659–4663

Reid KBM, Bentley DR, Campbell RD et al (1986) Complement system proteins which interact with C3b or C4b. Immunol Today 7:230

Richards A, Kemp EJ, Liszewski MK et al (2003) Mutations in human complement regulator, membrane cofactor protein (CD46), predispose to development of familial hemolytic uremic syndrome. Proc Natl Acad Sci U S A 100:12966–12971

Richards A, Kavanagh D, Atkinson JP (2007) Inherited complement regulatory protein deficiency predisposes to human disease in acute injury and chronic inflammatory states. Adv Immunol 97:141–177

Riley RC, Tannenbaum PL, Abbott DH, Atkinson JP (2002) Cutting edge: Inhibiting measles virus infection but promoting reproduction: an explanation for splicing and tissue-specific expression of CD46. J Immunol 169:5405–5409

Riley-Vargas RC, Atkinson JP (2003) Expression of membrane cofactor protein (MCP; CD46) on spermatozoa: just a complement inhibitor? Mod Aspects Immunobiol 3:75–78

Riley-Vargas RC, Gill DB, Kemper C, Liszewski MK, Atkinson JP (2004) CD46: expanding beyond complement regulation. Trends Immunol 25:496–503

Riley-Vargas RC, Lanzendorf S, Atkinson JP (2005) Targeted and restricted complement activation on acrosome-reacted spermatozoa. J Clin Invest 115:1241–1249

Rima BK, Earle JA, Baczko K et al (1997) Sequence divergence of measles virus haemagglutinin during natural evolution and adaptation to cell culture. J Gen Virol 78:97–106

Rivailler P, Trescol-Biemont MC, Gimenez C, Rabourdin-Combe C, Horvat B (1998) Enhanced MHC class II-restricted presentation of measles virus (MV) hemagglutinin in transgenic mice expressing human MV receptor CD46. Eur J Immunol 28:1301–1314

Roncarolo MG, Bacchetta R, Bordignon C, Narula S, Levings MK (2001) Type 1 regulatory cells. Immunol Rev 182:68–79

Russell SM, Sparrow RL, McKenzie IFC, Purcell DFJ (1992) Tissue-specific and allelic expression of the complement regulator CD46 is controlled by alternative splicing. Eur J Immunol 22:1513–1518

Sakaguchi S (2000) Regulatory T cells: key controllers of immunologic self-tolerance. Cell 101:455–458

Sanchez A, Feito MJ, Rojo JM (2004) CD46-mediated costimulation induces a Th1-biased response and enhances early TCR/CD3 signaling in human CD4$^+$ T lymphocytes. Eur J Immunol 34:2439–2448

Santiago C, Bjorling E, Stehle T, Casasnovas JM (2002) Distinct kinetics for binding of the CD46 and SLAM receptors to overlapping sites in the measles virus hemagglutinin protein. J Biol Chem 277:32294–32301

Santoro F, Kennedy PE, Locatelli G, Malnati MS, Berger EA, Lusso P (1999) CD46 is a cellular receptor for human herpesvirus 6. Cell 99:817–827

Santoro F, Greenstone HL, Insinga A et al (2003) Interaction of glycoprotein H of human herpesvirus 6 with the cellular receptor CD46. J Biol Chem 278:25964–25969

Schneider-Schaulies J, Dunster LM, Kobune F, Rima B, ter Meulen V (1995) Differential down-regulation of CD46 by measles virus strains. J Virol 69:7257–7259

Schnoor J-J, Dunster LM, Nanan R, Schneider-Schaulies J, Schneider-Schaulies S, ter Meulen V (1995) Measles virus-induced down-regulation of CD46 is associated with enhanced sensitivity to complement-mediated lysis of infected cells. Eur J Immunol 25:976–984

Segerman A, Atkinson JP, Marttila M, Dennerquist V, Wadell G, Arnberg N (2003) Adenovirus type 11 uses CD46 as a cellular receptor. J Virol 77:9183–9191

Seki F, Takeda M, Minagawa H, Yanagi Y (2006) Recombinant wild-type measles virus containing a single N481Y substitution in its haemagglutinin cannot use receptor CD46 as efficiently as that having the haemagglutinin of the Edmonston laboratory strain. J Gen Virol 87:1643–1648

Seya T, Turner JR, Atkinson JP (1986) Purification and characterization of a membrane protein (gp45–70) that is a cofactor for cleavage of C3b and C4b. J Exp Med 163:837–855

Seya T, Ballard LL, Bora NS, Kumar V, Cui W, Atkinson JP (1988) Distribution of membrane cofactor protein (MCP) of complement on human peripheral blood cells. An altered form is found on granulocytes. Eur J Immunol 18:1289–1294

Seya T, Hirano A, Matsumoto M, Nomura M, Ueda S (1999) Human membrane cofactor protein (MCP, CD46): multiple isoforms and functions. Int J Biochem Cell Biol 31:1255–1260

Shibahara K, Hotta H, Katayama Y, Homma M (1994) Increased binding activity of measles virus to monkey red blood cells after long-term passage in Vero cell cultures. J Gen Virol 75:3511–3516

Shingai M, Inoue N, Okuno T et al (2005) Wild-type measles virus infection in human CD46/CD150-transgenic mice: CD11c-positive dendritic cells establish systemic viral infection. J Immunol 175:3252–3261

Simpson KL, Holmes CH (1994) Presence of the complement-regulatory protein membrane cofactor protein (MCP, CD46) as a membrane-associated product in seminal plasma. J Reprod Fertil 102:419–424

Smith A, Santoro F, Di Lullo G, Dagna L, Verani A, Lusso P (2003) Selective suppression of IL-12 production by human herpesvirus 6. Blood 102:2877–2884

Takeuchi K, Takeda M, Miyajima N, Kobune F, Tanabayashi K, Tashiro M (2002) Recombinant wild-type and Edmonston strain measles viruses bearing heterologous H proteins: role of H protein in cell fusion and host cell specificity. J Virol 76:4891–4900

Tatsuo H, Ono N, Tanaka K, Yanagi Y (2000) SLAM (CDw150) is a cellular receptor for measles virus. Nature 406(6798):893–897

Thorley BR, Milland J, Christiansen D et al (1997) Transgenic expression of a CD46 (membrane cofactor protein) minigene: studies of xenotransplantation and measles virus infection. Eur J Immunol 27:726–734

Tsujimura A, Shida K, Kitamura M et al (1998) Molecular cloning of a murine homologue of membrane cofactor protein (CD46): preferential expression in testicular germ cells. Biochem J 330:163–168

van Binnendijk RS, van der Heijden RW, Osterhaus AD (1995) Monkeys in measles research. Curr Top Microb Immunol 191:135–148

Varior-Krishnan G, Trescol-Biemont MC, Naniche D, Rabourdin-Combe C, Gerlier D (1994) Glycosyl-phosphatidylinositol-anchored and transmembrane forms of CD46 display similar measles virus receptor properties: virus binding, fusion, and replication; down-regulation by hemagglutinin and virus uptake and endocytosis for antigen presentation by major histocompatibility complex class II molecules. J Virol 68:7891–7989

Vongpunsawad S, Oezgun N, Braun W, Cattaneo R (2004) Selectively receptor-blind measles viruses: identification of residues necessary for SLAM- or CD46-induced fusion and their localization on a new hemagglutinin structural model. J Virol 78:302–313

Walport MJ (2001) Complement. First of two parts. N Engl J Med 344:1058–1066

Walport MJ (2001) Complement. Second of two parts. N Engl J Med 344:1140–1144

Whaley K, Loos M, Weiler JM (1993) Complement in health and disease. 2nd edn. Kluwer Academic, London

Wild TF, Malvoisin E, Buckland R (1991) Measles virus: both the haemagglutinin and fusion glycoproteins are required for fusion. J Gen Virol 72:439–442

Wild TF, Fayolle J, Beauverger P, Buckland R (1994) Measles virus fusion: role of the cysteine-rich region of the fusion glycoprotein. J Virol 68:7546–7548

Wilton AN, Johnstone RW, McKenzie IF, Purcell DF (1992) Strong associations between RFLP and protein polymorphisms for CD46. Immunogenetics 36:79–85

Xu C, Mao D, Holers VM, Palanca B, Cheng AM, Molina H (2000) A critical role for the murine complement regulator Crry in fetomaternal tolerance. Science 287:498–501

Yanagi Y, Takeda M, Ohno S (2006) Measles virus: cellular receptors, tropism and pathogenesis. J Gen Virol 87:2767–2779

Yannoutsos N, Ijzermans JN, Harkes C et al (1996) A membrane cofactor protein transgenic mouse model for the study of discordant xenograft rejection. Genes Cell 1:409–419

Yu GH, Holers VM, Seya T, Ballard L, Atkinson JP (1986) Identification of a third component of complement-binding glycoprotein of human platelets. J Clin Invest 78:494–501

Zaffran Y, Destaing O, Roux A et al (2001) CD46/CD3 costimulation induces morphological changes of human T cells and activation of Vav, Rac, and extracellular signal-regulated kinase mitogen-activated protein kinase. J Immunol 167:6780–6785

Zheng XL, Sadler JE (2008) Pathogenesis of thrombotic microangiopathies. Annu Rev Pathol 3:249–277

Chapter 4
Measles Virus Glycoprotein Complex Assembly, Receptor Attachment, and Cell Entry

C.K. Navaratnarajah, V.H.J. Leonard, and R. Cattaneo(✉)

Contents

Introduction	60
Glycoprotein Complex Assembly	60
The Attachment Protein Hemagglutinin Dimer	62
The Fusion Protein Trimer	63
Hetero-oligomerization of F and H Contributes to Particle Assembly	64
Receptor Attachment, Membrane Fusion, and Cell Entry	65
Cell Entry Through CD46: Scaffolding Fusion Through Receptor-H Protein Interactions	66
Cell Entry Through SLAM: Dynamic Interactions and Conformational Changes	68
Cell Entry Through EpR: The Footprint	69
Cell Entry Through Designated Receptors: A Compliant Membrane Fusion System	70
Model for MV H-Mediated Fusion Activation	70
Perspectives	72
References	73

Abstract Measles virus (MV) enters cells by membrane fusion at the cell surface at neutral pH. Two glycoproteins mediate this process: the hemagglutinin (H) and fusion (F) proteins. The H-protein binds to receptors, while the F-protein mediates fusion of the viral and cellular membranes. H naturally interacts with at least three different receptors. The wild-type virus primarily uses the signaling lymphocyte activation molecule (SLAM, CD150) expressed on certain lymphatic cells, while the vaccine strain has gained the ability to also use the ubiquitous membrane cofactor protein (MCP, CD46), a regulator of complement activation. Additionally, MV infects polarized epithelial cells through an unidentified receptor (EpR). The footprints of the three receptors on H have been characterized, and the focus of research is shifting to the characterization of receptor-specific conformational changes that occur in the H-protein dimer and how these are transmitted to the F-protein trimer. It was also shown that MV attachment and cell entry can be readily targeted to designated receptors by adding specificity determinants to the H-protein. These

---

R. Cattaneo
Department of Molecular Medicine, Virology and Gene Therapy Graduate Track, Mayo Clinic College of Medicine, 200 1st St SW, Guggenheim 1838, Rochester, MN 55905, USA, e-mail: Cattaneo.Roberto@mayo.edu

studies have contributed to our understanding of membrane fusion by the glycoprotein complex of paramyxoviruses in general.

Introduction

This review focuses on the measles virus (MV) glycoprotein complex: we discuss its assembly and how its interactions with different receptors result in cell entry. The MV glycoprotein complex makes the first contact with the host, targeting the infection to specific cells and thus governing tropism and pathogenesis. MV, one of the most contagious human pathogens, is transmitted by aerosols, infecting a new host via the upper respiratory tract (Panum 1939). It has been assumed that MV infects the respiratory epithelium from the luminal side before spreading in lymphatic cells (Griffin 2007; Cherry 2003). However, the identification of the signaling lymphocytic activation molecule (SLAM, CD150) as the primary MV receptor (Tatsuo et al. 2000) and recent work with selectively receptor-blind MV and animal morbilliviruses (Leonard et al. 2008; von Messling et al. 2006) have brought compelling evidence for a new model of MV dissemination postulating that the systemic spread of wild-type MV depends only on infection of SLAM-expressing lymphatic cells, without initial virus amplification in the respiratory epithelium (Fig. 1) (Leonard et al. 2008; de Swart et al. 2007; von Messling et al. 2006; Yanagi et al. 2006). This model implies that MV does not cross the respiratory epithelium immediately after contagion, but only when it leaves the host (Fig. 1). Moreover, it predicts that a virus that does not recognize the unidentified epithelial receptor (EpR) will not be shed. This prediction was confirmed by generating a recombinant EpR-blind MV and demonstrating that it spread in lymphocytes and remained virulent in rhesus monkeys, but importantly was not shed in the respiratory tract (Leonard et al. 2008). The protein initially identified as the vaccine (Edmonston) strain receptor, the membrane cofactor protein (MCP, CD46), a ubiquitous regulator of complement activation, seems to be of minor relevance for wild type MV infections (see the chapter by Y. Yanagi et al., this volume), and the ability of the vaccine strain to use CD46 partially explains its attenuated phenotype (Condack et al. 2007; Schneider-Schaulies et al. 1995).

Glycoprotein Complex Assembly

Understanding how the glycoprotein complex is assembled is important for the characterization of the subsequent mechanisms of receptor-binding and cell entry. The MV envelope covers particles ranging in diameter from 120–300 nm (Armstrong et al. 1982; Casali et al. 1981). The envelope is traversed by the hemagglutinin (H) and fusion (F) glycoproteins, which form spikes that extend about 10–15 nm from the surface of the membrane (Fig. 2). It surrounds the nucleocapsid core, which

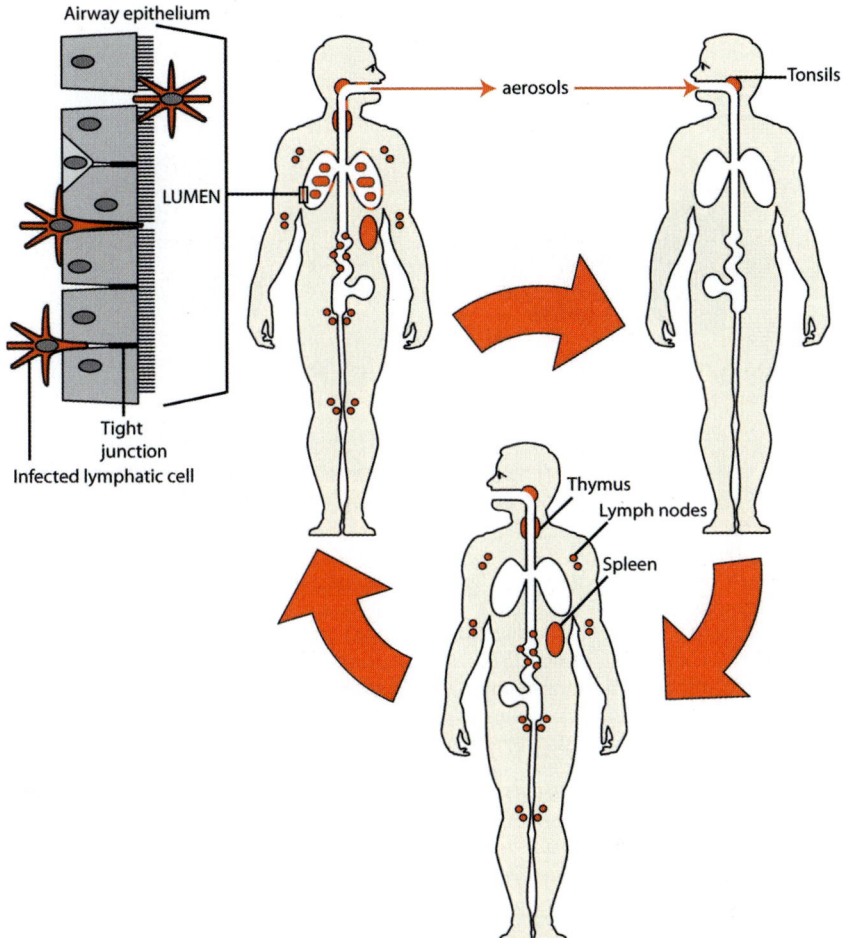

Fig. 1 Disease mechanism of MV. MV is transmitted by aerosol and respiratory secretions. Primary infection may start in SLAM-expressing lymphatic cells in the tonsils (*top right*) and rapidly disseminate to all lymphatic organs (*bottom*). Destabilization of the respiratory epithelium through infected lymphatic cells contacting EpR (*inset, top left*) may result in epithelial crossing of these cells, coughing, and contagion (*top center*)

typically includes several genomes (Rager et al. 2002) tightly encapsidated by a helically arranged nucleocapsid (N) protein. Also associated with the genome are two other proteins, the polymerase (L) and a polymerase cofactor (phosphoprotein, P), all of which together form a replicationally active ribonucleoprotein (RNP) complex (see the chapter by B.K. Rima and W.P. Duprex, this volume). The MV RNP is condensed by the matrix (M) protein (Cathomen et al. 1998), which is hydrophobic and interacts with cell membranes, forming leaflet structures at the inner side of the viral envelope (Fig. 2, bottom). It may mediate the contacts between the cytoplasmic tails of the F and H glycoproteins and the RNP complex.

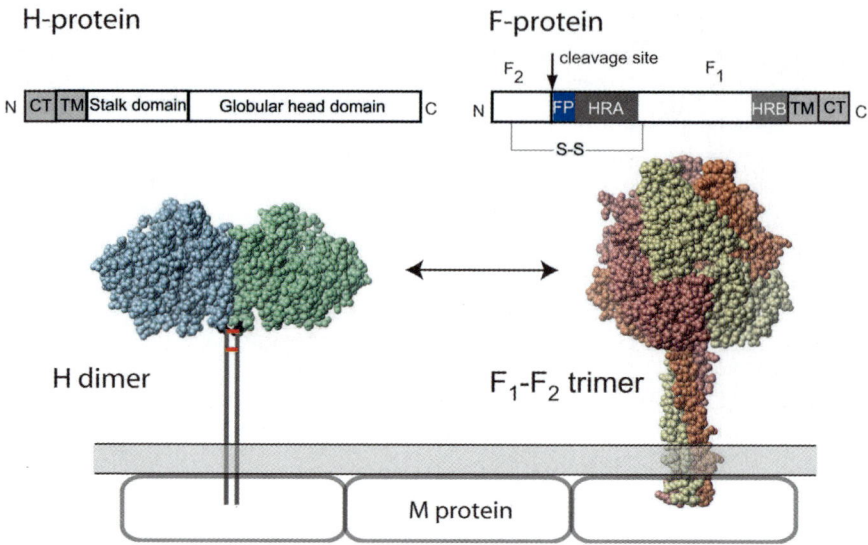

Fig. 2 The MV envelope glycoproteins. *Top left:* Schematic of the MV H-protein. *CT*, cytoplasmic tail; *TM*, transmembrane segment; *N* and *C*, N- and C-terminus, respectively. *Top right:* Schematic of the MV F-protein. The disulfide bond (S-S) holding cleavage fragments F_1 and F_2 is indicated. *FP*, fusion peptide; *HRA*, *HRB*, heptad repeats A and B, respectively. *Center and bottom:* Quaternary structure and interactions of the H- and F-proteins. A space-filling representation of the crystal structure of the MV H dimer (Hashiguchi et al. 2007) and the PIV5 F trimer (Yin et al. 2006) is shown. PIV5 is a paramyxovirus related to MV. The stalk, TM, and CT regions of the H-protein are represented by *two vertical lines* with the two disulfide bonds that hold the H dimer together represented by *horizontal red lines*. The two monomers of the H-protein dimer and the three monomers of the F-protein trimer are shown in different colors for clarity. The interactions of the H- and F-proteins (*double-headed arrow*) are thought to be mediated via the head and stalk domains. The M protein (*clear boxes*) interacts with the membrane (*gray box*) and the CTs of the H- and F-proteins

Here we discuss first the biosynthesis and intercellular transport of the H-protein dimer and F-protein trimer and then the interactions of the H and F envelope oligomers leading to the assembly of fusion-competent glycoprotein complexes.

The Attachment Protein Hemagglutinin Dimer

Receptor attachment of MV and the other morbilliviruses is mediated by the H-protein that has, in addition, a fusion-support function: co-expression of MV H with F is required for fusion (Cattaneo and Rose 1993; Wild et al. 1991). The H-protein binding to SLAM, or other receptors, provides the activation energy to trigger the F-protein to carry out membrane fusion (Navaratnarajah et al. 2008; Yin et al. 2006). Unlike other members of the *Paramyxoviridae*, the morbillivirus H-proteins do not have neuraminidase activity, since H does not bind sialic acid (Hashiguchi et al. 2007).

MV H is a 617 amino acid (78-kDa) type II transmembrane glycoprotein, which is comprised of an N-terminal cytoplasmic tail, a membrane-spanning domain, and an extracellular membrane-proximal stalk region connected to a large C-terminal globular head (Fig. 2, linear map on the top left) (Alkhatib and Briedis 1986). Receptor-binding residues have been mapped to this head domain (Leonard et al. 2008; Navaratnarajah et al. 2008; Tahara et al. 2008; Masse et al. 2004; Vongpunsawad et al. 2004; Hsu et al. 1998). The hemagglutinin-neuraminidase (HN) receptor attachment proteins of other paramyxoviruses exist as tetramers at the viral surface, consisting of a dimer of two disulfide linked homodimers (Lamb 1993). However, the full length MV H only forms dimers (Plemper et al. 2000), and the head domain of H was also crystallized as a dimer (Hashiguchi et al. 2007). We thus refer to the H oligomeric form simply as a dimer. Two disulfide bonds are formed by cysteine residues (C139 and C154) located at the C-terminal end of the stalk domain, just below the globular head (Fig. 1, center, red lines) (Hashiguchi et al. 2007; Plemper et al. 2000).

The crystal structure of the MV H-protein globular head domain was recently reported by two groups (Colf et al. 2007; Hashiguchi et al. 2007). Hashiguchi et al. (2007) reported a dimeric structure (Fig. 2, center left) while Colf et al. (2007) published a monomeric structure. The head domain of MV H (residues 154-607) exhibits a β-propeller structure comprised of six anti-parallel β-sheets (β-sheets 1–6) organized as a superbarrel. This structure most closely resembles the human parainfluenza virus 3 (hPIV3) HN crystal structure (Colf et al. 2007). The overall fold is similar to the sialidase fold previously determined for the HN proteins of hPIV3, Newcastle disease virus (NDV) and parainfluenza virus 5 (PIV5, formerly known as Simian virus 5) (Yuan et al. 2005; Lawrence et al. 2004; Crennell et al. 2000). However, the 3.9-Å root mean square deviation between the Cα atoms of MV H and hPIV3 HN indicates that the H-protein exhibits considerable structural divergence from these proteins. While other paramyxovirus attachment proteins are globular, MV H-protein exhibits a cube-shaped structure. A significant area of the H-protein is covered with N-linked sugars (attached to N200 and N215) and is thus unavailable for receptor interaction (Hashiguchi et al. 2007). Also apparent from the crystal structure is that, in contrast to other paramyxovirus attachment proteins, the two H-protein molecules making up the homodimer are highly tilted with respect to each other. This becomes important when discussing the location of residues implicated in receptor-specific fusion support (see below) (Navaratnarajah et al. 2008).

The Fusion Protein Trimer

The MV F-protein is a 553 amino acid type I transmembrane glycoprotein (Fig. 2, linear map on the top right) (Richardson et al. 1986). A cleavable 28-residue signal sequence at the N-terminus of the nascent polypeptide directs it to the endoplasmic reticulum. A transmembrane (TM) domain near the C-terminal end anchors it in the membrane, leaving a short 33 amino acid cytoplasmic tail. The F-protein is synthesized as a precursor polypeptide (F_0, 60 kDa), which trimerizes in the endoplasmic reticulum (Plemper et al. 2001). F_0 is then cleaved by the ubiquitous intracellular protease furin in the trans-Golgi (Bolt and Pedersen 1998; Watanabe et al. 1995). Proteolytic cleavage results in a metastable F-protein (Fig. 2, center right) that consists of a membrane-spanning (F_1, 40 kDa), and a membrane-distal subunit (F_2, 20 kDa), that is assembled into virus particles (Lamb 1993). The large F_1 fragment is anchored to the membrane via the TM domain, while the small F_2 fragment is covalently linked to F_1 by a disulfide bond (Fig. 2, top right). The new N-terminus of the F_1 fragment contains a hydrophobic stretch of amino acids, which comprises the fusion peptide (FP) that is inserted in to the target membrane during fusion. The sequences adjacent to the FP and TM exhibit a 4–3 (heptad) pattern of hydrophobic repeats and are named HRA and HRB, respectively (Fig. 2, top right). HRA and HRB sequences are separated by approximately 250 residues and mutagenesis of the heptad repeats adversely affects fusion (Buckland et al. 1992). The MV F-protein has three N-linked carbohydrate chains, all located in the F_2 subunit, that are necessary for proper proteolytic processing and transport to the cell surface (Hu et al. 1995).

The MV F-protein is classified as a class I fusion protein (Kielian and Rey 2006; Yin et al. 2006). The fusion proteins of retroviruses, coronaviruses, Ebola virus, influenza virus, as well as other paramyxoviruses also belong to this class. Class I fusion proteins mediate membrane fusion by coupling irreversible protein refolding to membrane juxtaposition (Lamb and Parks 2007). This is accomplished by discrete conformational changes of a metastable F-protein structure to a lower energy structure. The F-protein found in virions is referred to as the pre-fusion structure and, as it mediates membrane fusion, the F-protein adopts the post-fusion form. The cleavage of F_0 into F_1–F_2 primes the protein for membrane fusion. Activation of the F-protein results in the insertion of the FP, located in the F_1 subunit, into the target membrane. This is followed by dramatic conformational rearrangements of the F trimer, which result in a transient hairpin intermediate and the subsequent formation of a stable six-helix bundle (6HB) structure in which the HRA peptides form a central three-stranded coiled coil, and the HRB peptides pack in an antiparallel manner into hydrophobic grooves on the coiled-coil surface. As a result of these conformational changes, the FP and TM domain, and thus the target and donor membranes, are now in close proximity to each other (Yin et al. 2006). This eventually leads to the formation of a fusion pore through which the RNP complex can enter the cell.

Hetero-oligomerization of F and H Contributes to Particle Assembly

The M-protein is considered the assembly organizer of paramyxovirus particles, and it was formally shown for MV that M regulates the fusion efficiency of the glycoprotein complex (Cathomen et al. 1998). The M-protein assembles in lattice-like structures at the inner side of the plasma membrane and binds to the cytoplasmic tails of the glycoproteins (Fig. 1, bottom) (Buechi and Bachi 1982), bridging the envelope to the ribonucleocapsid.

Whereas early models of MV assembly assumed that the H- and F-protein ectodomains interact only at the cell surface, it was then shown that the H- and F-proteins form strong lateral interactions already in the endoplasmic reticulum (Plemper et al. 2001). This is in contrast to other paramyxoviruses such as hPIV3 and PIV5, where only minimal amounts of intracellular complexes between HN and F are detected (Corey and Iorio 2007). The H and F glycoprotein complexes are transported through the secretory pathway to the plasma membrane, where they will sustain cell-to-cell fusion or virus particle release. While the exact interaction sites and stoichiometry of H and F in these glycoprotein complexes is yet to be determined, based on the fusion-support capacity of chimeric attachment proteins from different paramyxoviruses and antibody mapping studies, it appears that the stalk and head domains of H are responsible for conferring F specificity (Fournier et al. 1997).

A key question that remains to be addressed in the MV entry process, and indeed for all paramyxoviruses, is how the receptor attachment protein triggers the F conformational changes upon receptor binding. Plemper et al. (2002) observed that the stability of the H-F complex was a modulator of virus-induced cell-to-cell fusion and that destabilizing the H–F interaction results in a significant increase in lateral cell-to-cell fusion. This observation is in agreement with a model proposed for NDV fusion activation, where the pre-fusion state of the F-protein is stabilized by association with the HN-protein and upon receptor attachment the HN–F complex dissociates, thus leading to F activation and fusion (McGinnes and Morrison 2006). Recent results of Mühlebach et al. (2008) highlight the importance of the $H–(F_1+F_2)$ glycoprotein complex for fusion. Mutations that reduced the amount of this complex were less fusogenic, while mutations that increased its availability were more fusogenic (Mühlebach et al. 2008). All these observations are consistent with a role for the H-protein in stabilizing the F-protein prior to fusion. The strong H–F interaction observed in infected cells may be responsible for preventing premature fusion after furin cleavage of the F-protein has converted it to a metastable pre-fusion structure.

Receptor Attachment, Membrane Fusion, and Cell Entry

This section presents a discussion of the H-protein interactions with the two known receptors (CD46 and SLAM) and the putative epithelial cell receptor (EpR). Experiments with CD46 have revealed the importance of a receptor-H protein scaffold as a prerequisite for fusion. Subsequent experiments with SLAM have demonstrated that there are receptor-specific H-protein conformational changes that lead to fusion activation. The footprint of EpR has already been defined on H, even if the identity of this receptor remains unknown. Finally, the ability to retarget

MV to specific receptors via specificity determinants fused to the C-terminus of the H-protein is discussed.

Cell Entry Through CD46: Scaffolding Fusion Through Receptor-H Protein Interactions

The first MV receptor to be identified was the ubiquitous regulator of complement activation, CD46 (see the chapter by C. Kemper and J.P. Atkinson et al., this volume) (Dorig et al. 1993; Naniche et al. 1993). MV adapts to use CD46 by accumulated changes in the H-protein, the most significant of which is an N481Y substitution (Bartz et al. 1996; Lecouturier et al. 1996). The attenuated phenotype and altered tropism of the MV vaccine strain compared to the wild-type strains are partly explained by the use of CD46. Although CD46 is not the principal cellular receptor for the wild-type strain of MV, experiments with this receptor have yielded important insights into the mechanism of fusion activation in MV and other morbilliviruses (Buchholz et al. 1996, 1997).

These experiments provided insights into how the H-protein and a receptor may hold the virus in place before the F-protein trimer unfolds and mediates membrane fusion. Buchholz et al. (1996) demonstrated that increasing the length of the CD46 protein, thereby effectively increasing the distance between the viral and cell membranes, enhances binding but reduces fusion. The increase in virus binding can be explained by the alleviation of steric hindrance as the attachment site on the receptor is moved further away from the cell surface. The fact that the fusion efficiency decreased as a result of this indicates that the viral and cellular membranes need to be at a certain distance from each other for fusion to occur. This distance is probably defined by the need for the fusion peptide to efficiently insert into the target cell membrane. Another key finding was that the CD46 receptors with increased length had a dominant-negative effect on fusion when co-expressed with functional CD46 receptors, even at an unfavorable molar ratio. This is indirect evidence for the existence of a MV fusion complex in which multiple H-protein dimers bind CD46 and form a scaffold surrounding multiple F-protein trimers. Co-expressing CD46 molecules of different length would result in an irregular scaffold that would be unable to support fusion, thus explaining the dominant-negative effect.

The extracellular domain of CD46 consists of four complement control protein domains (CCP1–4) (Fig. 3D, top left). The MV binding site has been localized to the membrane distal CCP1 and CCP2 domains (Casasnovas et al. 1999; Devaux et al. 1997; Manchester et al. 1997). Specifically, the region comprising CD46 residues 37–59 is involved in H-protein interactions (Buchholz et al. 1997) and mutation of R59 in the CCP1 domain interferes with viral fusion and CD46 downregulation (Hsu et al. 1999). On the H-protein, V451 and Y481 have been identified as CD46-dependent fusion-support residues (Lecouturier et al. 1996). Binding studies with the soluble extracellular domain of the H-protein and soluble receptors demonstrated the importance of the tyrosine at position 481 (Navaratnarajah

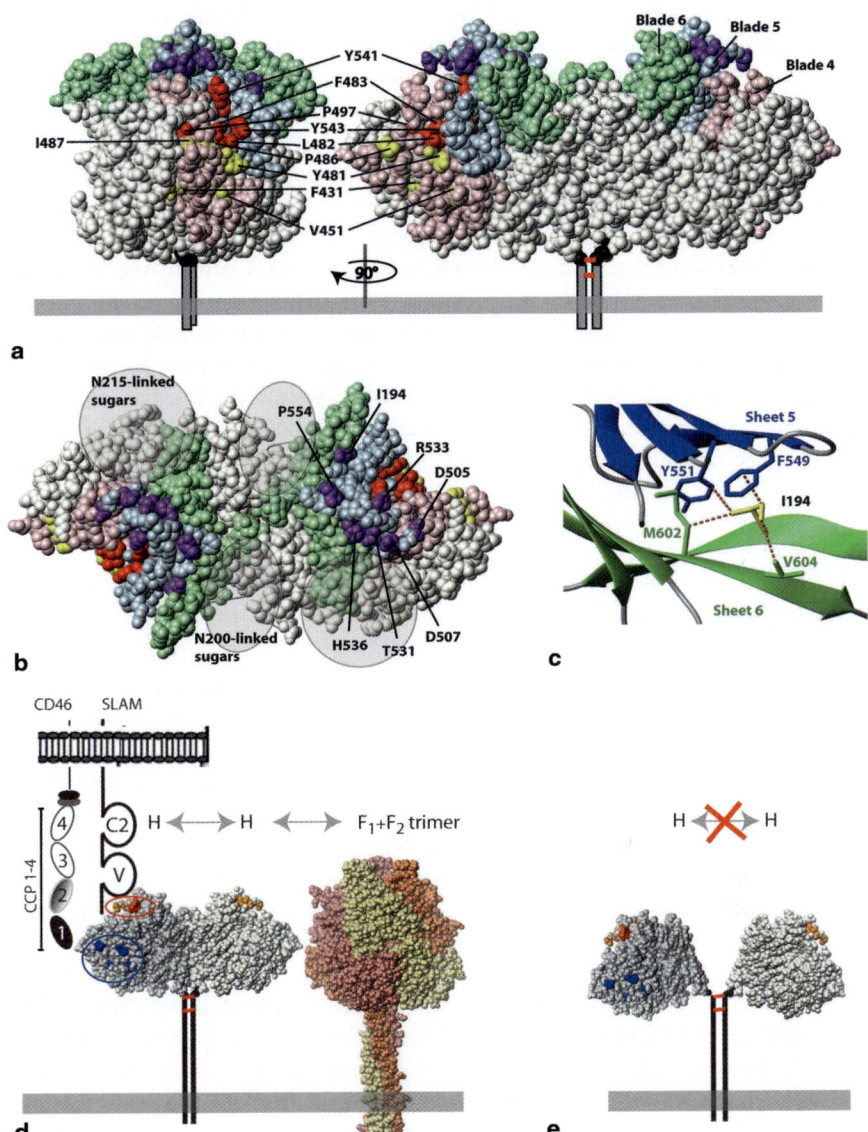

Fig. 3 a–e Receptor footprints on the H-protein dimer and a model for fusion activation. **A** Side view of the MV H-protein homodimer crystal structure depicted as a space-filling model. Residues 1–153, not present in the crystal structure and comprising the cytoplasmic tail, transmembrane, and stalk regions are represented as *vertical gray boxes*. The *two red lines* represent the C154-C154 (residues colored *black*) and C139-C139 disulfide bonds that link the two monomers in the homodimer. The membrane is illustrated as a *horizontal gray box*. β-Propeller blades 4, 5, and 6 are surface shaded *pink*, *blue*, and *green*, respectively. SLAM-, CD46-, and EpR-specific residues are shaded *purple*, *yellow*, and *red*, respectively. **B** Top view of the H homodimer generated by rotating the view depicted in **A** (*right panel*) 90° around the x-axis. The regions occluded by the N-linked oligosaccharide chains are indicated by *gray ovals*. The color coding is identical to **A**.

et al. 2008; Hashiguchi et al. 2007). The wild-type strain of MV carries an asparagine at this position and does not interact with CD46 in surface plasmon resonance-based binding assays. However, when the asparagine was substituted by tyrosine, the soluble H-protein showed an appreciable interaction with CD46.

Figure 3 presents the recently published crystal structure of the H dimer with receptor-specific residues indicated (Navaratnarajah et al. 2008; Hashiguchi et al. 2007). In this structure, the two H monomers tilt oppositely toward the horizontal plane (Fig. 3A, right panel). This orients propeller blades 5 (blue surface) and 6 (green surface) with the SLAM-relevant residues upward toward the target cell (Fig. 3A and 3B). Mapping of the CD46-specific fusion-support residues (F431, V451, Y481, P486, and I487) on to the MV H-protein structure defines a region that involves propeller blade 4 (pink surface) located on the side of the H-dimer (Fig. 3A and 3B, residues shaded yellow). Most CD46-relevant residues are located in the bottom half of the dimer. This site is adjacent to but distinct from the SLAM- and EpR-specific sites (see below).

Cell Entry Through SLAM: Dynamic Interactions and Conformational Changes

SLAM is the primary cellular receptor for the wild-type strains (Erlenhoefer et al. 2001; Hsu et al. 2001; Ono et al. 2001a; Tatsuo et al. 2000). Its tissue-specific expression is consistent with MV lymphotropism and may be the fundament for pathogenesis and immunosuppression (see the chapter by Y. Yanagi et al. this volume, and chapter by S. Schneider-Schaulies and J. Schneider-Schaulies, accompanying volume). SLAM is composed of two immunoglobulin superfamily domains, V and C2, followed by a TM and a cytoplasmic domain (Fig. 3D, top left) (Cocks et al. 1995). The V domain interacts with H (Ono et al. 2001b). Specifically, I60, H61, and V63 are key residues supporting this interaction, forming a putative MV H binding site (Ohno et al. 2003).

Mutagenesis of the H-protein based on structural models has characterized residues I194, D505, D507, Y529, D530, T531, R533, H536, Y553, and P554 as important for SLAM-dependent fusion (Navaratnarajah et al. 2008; Masse et al. 2004; Vongpunsawad et al. 2004). Furthermore, it was shown that mutating

---

Fig. 3 (Continued) **C** Interactions of I194 with neighboring MV H residues. Closest distances between heavy atoms of I194 and interaction partners are depicted as *dotted red lines*. I194 has van der Waals contacts with Y551 and F549 on sheet 5 (*blue*) and with M602 and V604 on sheet 6 (*green*). **D** Close apposition of H dimers and F trimers before fusion activation. Schematics of the SLAM and CD46 receptors are illustrated next to their respective binding sites on the H-protein (*red circle*, SLAM; *blue circle*, CD46). Complement control protein domains 1–4 (CCP 1–4) of CD46 and immunoglobulin domains V and C2 of SLAM are indicated. Prior to receptor binding, the H dimer is closely associated with the F trimer. **E** H-protein dimer destabilization after receptor-binding. Conformational changes of the H dimer may disrupt the H-(F_1+F_2) interaction, resulting in fusion activation

I194 ablates SLAM-binding (Navaratnarajah et al. 2008). In contrast, mutating Y529, D530, R533, and P553 (termed the β-sheet 5 quartet), while interfering with SLAM-dependent fusion, does not affect SLAM binding. The location of these residues is visualized in Fig. 3 on the crystal structure of the H-protein ectodomain dimer. The SLAM relevant residues (purple) are centrally located on the top of the H-protein homodimer.

The results of assays measuring SLAM-dependent fusion-support and SLAM-binding gave insights in to H-SLAM binding dynamics, which may be based on sequential H-protein conformational changes. We know that the initial SLAM interaction is influenced by I194, and Fig. 3C illustrates how this residue, located on one β-strand of propeller blade 6 (green), engages in van der Waals interactions (red dashed lines) with M602 and V604 on another β-strand of propeller blade 6, as well as with F549 and Y551 on a β-strand of propeller blade 5 (blue), stabilizing this interface. Thus, I194 may maintain a conformation of the H-protein conducive to SLAM-binding, rather than directly interacting with the receptor. The fact that four of the residues (β-sheet 5 quartet) are essential for SLAM-dependent fusion but not for SLAM binding proves that H-protein conformational changes can be receptor-specific (Navaratnarajah et al. 2008). It also implies that alternative pathways of H-protein conformational changes can converge to trigger F-protein unfolding and membrane fusion.

Cell Entry Through EpR: The Footprint

Lack of SLAM expression in epithelial cells suggests that another receptor exists (de Swart et al. 2007; von Messling et al. 2006; Yanagi et al. 2006). While the identity of the putative epithelial cell receptor (EpR) has remained elusive, two groups have independently mapped H-protein residues specifically sustaining EpR-dependent fusion (Leonard et al. 2008; Tahara et al. 2008). These studies identified H-protein residues L482, F483, P497, Y541, and Y543 (Fig. 3, red residues) located in an area distinct from that defined by the SLAM-interacting residues. They define a nonpolar valley running between propeller blades 4 and 5 (Fig. 3A, left panel). The nonpolar side chains of L482, F483, and P497, together with the uncharged polar side chains of Y541 and Y543, flank this valley situated between the SLAM- and the CD46-binding sites. The hydroxyl group of tyrosine 543 tops one of the ridges. Thus, uncharged polar and nonpolar residues govern the H–EpR interaction. It is not known which EpR-relevant residues are directly involved in receptor binding. Residues L482, F483, P497, Y541, and Y543 are 13%, 14%, 6%, 28%, and 17% solvent-exposed, respectively. Thus, the minimally exposed proline 497 may conduct a receptor-dependent conformational change. As for tyrosine 543, the hydroxyl group on the aromatic ring may play a central role in EpR binding: replacing it with phenylalanine, which differs from tyrosine by carrying a hydrogen atom in place of the hydroxyl group at this position, resulted in loss of EpR-specific fusion-support while maintaining SLAM-dependent function (Leonard et al. 2008).

Cell Entry Through Designated Receptors: A Compliant Membrane Fusion System

MV can be readily targeted to enter cells through a designated receptor by adding a specificity determinant to the C-terminus of the H-protein. First demonstrated for the epidermal growth factor receptor (Schneider et al. 2000), MV has since been targeted to a variety of different cell surface molecules (see the chapter by S. Russell and K.W. Peng, accompanying volume). Table 1 shows that receptor choice is not a limitation for membrane fusion. MV has been modified to utilize a broad range of cell surface antigens for cell entry, irrespective of their structure or function. These proteins can span the membrane only once, in type I or type II orientations, or several times, or even be anchored to the membrane by a glycosphingolipid. They can be monomeric, homo-oligomeric, or hetero-oligomeric, and have a wide range of functions, as well as cell-type specificity of expression. Equally important is the ability of the virus to accommodate the corresponding ligands, which range from the relatively small epidermal growth factor ligand to the significantly larger single-chain antibodies.

The ability to retarget MV entry, including detargeting of the natural receptors, coupled with its inherent oncolytic nature, has provided new perspectives for the development of MV-based phase I or phase I/II clinical trials of ovarian cancer, glioma, and myeloma (Liu et al. 2007) (see the chapter by S. Russell and K.W. Peng, accompanying volume, for an in-depth analysis). Two other developments enhancing specificity and efficacy of oncolytic MV are the use of cancer-specific proteases to activate the F-protein (Springfeld et al. 2006) and arming of the virus by prodrug convertases (Ungerechts et al. 2007a, 2007b). By altering the location of the furin cleavage site of F through addition of hexameric sequences recognized by matrix metalloproteinases, it was possible to generate a recombinant virus that only spreads in cancer cells expressing these proteases. Moreover, arming of MV by prodrug convertases sustains oncolytic efficacy even in mouse tumors set in fully immunocompetent mice (Ungerechts et al. 2007b). In particular, arming MV with the prodrug convertase purine nucleoside phosphorylase (PNP) is being sought in the context of non-Hodgkin lymphoma treatment, because PNP converts fludarabine phosphate, a drug used in combination with other chemotherapeutics to treat lymphoma (Ungerechts et al. 2007a). Finally, MV envelope proteins have been used to target retroviruses and lentiviruses, further extending the therapeutic potential of these vectors (Funke et al. 2008).

Model for MV H-Mediated Fusion Activation

Based on the data discussed above, the following model of MV-induced membrane fusion can be postulated. As shown in Fig. 3D, H dimers are associated with F_1–F_2 trimers prior to receptor attachment (Plemper et al. 2001). First, upon receptor binding the H-protein dimer creates a scaffold with the receptor (Fig. 3D, left), locating

Table 1 Measles virus natural and designated receptors

Synonyms	Predominant expression site	Size (kDa)	Transmembrane organization	Quaternary structure	Function	Reference
Natural receptors						
CD150 (SLAM)	Lymphatic cells	70–95	Type I	Monomer	Adhesion/signaling	Bucheit et al. 2003
CD46 (MCP)	Ubiquitous	55–65	Type I	Monomer	Regulator of complement activation	Peng et al. 2003
Designated receptors						
CD20	B lymphocytes	34	Tetraspanin	Homo-oligomers	Signaling/cell activation	Hammond et al. 2001
CD38	Leukocytes	45	Type II	Homodimer	ADP-ribosyl cyclase/signaling	
CEA	Early development transformed cells	180	GPI-anchor	–	Adhesion	
EGFR (ErbB-1, HER1)	Lung, GI tract, breast epithelium	170	Type I	Monomer	Signaling	Nakamura et al. 2004
EGFRvIII	Tumor cells	140	Type I	Monomer	Signaling	Nakamura et al. 2005
IGFR	Ubiquitous	130/95	Type I (β chain)	Heterotetramer	Signaling	Schneider et al. 2000
Intergrin αvβ3	OEFMP[a]	160/85	Type I/type I	Heterodimer	Adhesion/signaling/internalization	Hallak et al. 2005

[a]Osteoclasts, endothelial cells, fibroblasts, macrophages, platelets

the F trimer at an appropriate distance for the fusion peptide to reach the target membrane (Buchholz et al. 1996). Second, the H-receptor interaction triggers F to initiate the fusion process. As shown in Fig. 3E, the trigger may be simply the destabilization of the H-(F_1+F_2) interaction or it may entail a more complex set of H-protein conformational changes. As has been suggested for NDV (Ludwig et al. 2008; McGinnes and Morrison 2006), the interaction of the H-protein with F_1+F_2 may be essential for stabilizing the metastable pre-fusion form of the F-protein, thus preventing premature fusion (Plemper et al. 2002). For PIV5, it has been suggested that the interaction of the tetrameric attachment protein with sialic acid receptors destabilizes the tetramer, which in turn leads to changes in the interaction with F, which triggers F to mediate membrane fusion (Yuan et al. 2005). An analogous model based on dimer destabilization can be adopted for MV fusion: the interaction of MV H with its receptor may destabilize the H dimer (Fig. 3E), which in turn will destabilize the H–(F_1+F_2) interaction, resulting in F activation. In retrospect, the fact that many different specificity domains displayed on the H-protein can elicit fusion through targeted receptors can be rationalized by the availability of alternative pathways of H-protein conformational changes eliciting membrane fusion (Navaratnarajah et al. 2008).

Perspectives

In the last 15 years, the identification of two MV receptors, together with mutagenesis studies based on structural models of the MV H-protein, have supported the characterization of the H–receptor interactions that govern tropism and pathogenesis. The recently determined crystal structure of MV H allows accurate planning of the next phase of experimentation, which aims at understanding the molecular dynamics of receptor binding and fusion activation. More crystal structures of individual molecules or quaternary complexes are needed. However, these structures will represent only one possible conformation, while proteins and protein complexes are constantly moving in solution as a result of thermal energy. Further elucidation of the mechanism of fusion activation and cell entry will thus depend on our ability to study conformational changes through conformation-specific antibodies, sophisticated fusion assays, and computational modeling techniques based on molecular dynamic simulations.

In another new avenue of research, the insights gained by the mutagenesis studies on the determinants of MV tropism in the H-protein have yielded selectively receptor-blind viruses. These new tools have been used for basic research studies on the mechanisms of virulence and pathogenesis, and will be integrated in the next generation of oncolytic virotherapy clinical protocols. While measles is under control and may soon be eradicated, the study of MV biology remains of central importance for the development of MV-based replicating therapeutics to treat cancer, and for the generation of multivalent vaccines (see the chapter by M. Billeter and S. Udem, this volume).

References

Alkhatib G, Briedis DJ (1986) The predicted primary structure of the measles virus hemagglutinin. Virology 150:479–490

Armstrong MA, Fraser KB, Dermott E, Shirodaria PV (1982) Immunoelectron microscopic studies on haemagglutinin and haemolysin of measles virus in infected HEp2 cells. J Gen Virol 59:187–192

Bartz R, Brinckmann U, Dunster LM, Rima B, ter Meulen V, Schneider-Schaulies J (1996) Mapping amino acids of the measles virus hemagglutinin responsible for receptor (CD46) downregulation. Virology 224:334–337

Bolt G, Pedersen IR (1998) The role of subtilisin-like proprotein convertases for cleavage of the measles virus fusion glycoprotein in different cell types. Virology 252:387–398

Bucheit AD, Kumar S, Grote DM, Lin Y, von Messling V, Cattaneo RB, Fielding AK (2003) An oncolytic measles virus engineered to enter cells through the CD20 antigen. Mol Ther 7:62–72

Buchholz CJ, Schneider U, Devaux P, Gerlier D, Cattaneo R (1996) Cell entry by measles virus: long hybrid receptors uncouple binding from membrane fusion. J Virol 70:3716–3723

Buchholz CJ, Koller D, Devaux P, Mumenthaler C, Schneider-Schaulies J, Braun W, et al (1997) Mapping of the primary binding site of measles virus to its receptor CD46. J Biol Chem 272:22072–22079

Buckland R, Malvoisin E, Beauverger P, Wild F (1992) A leucine zipper structure present in the measles virus fusion protein is not required for its tetramerization but is essential for fusion. J Gen Virol 73:1703–1707

Buechi M, Bachi T (1982) Microscopy of internal structures of Sendai virus associated with the cytoplasmic surface of host membranes. Virology 120:349–359

Casali P, Sissons JG, Fujinami RS, Oldstone MB (1981) Purification of measles virus glycoproteins and their integration into artificial lipid membranes. J Gen Virol 54:161–171

Casasnovas JM, Larvie M, Stehle T (1999) Crystal structure of two CD46 domains reveals an extended measles virus-binding surface. EMBO J 18:2911–2922

Cathomen T, Mrkic B, Spehner D, Drillien R, Naef R, Pavlovic J, et al (1998) A matrix-less measles virus is infectious and elicits extensive cell fusion: consequences for propagation in the brain. EMBO J 17:3899–3908

Cattaneo R, Rose JK (1993) Cell fusion by the envelope glycoproteins of persistent measles viruses which caused lethal human brain disease. J Virol 67:1493–1502

Cherry J (2003) Measles virus. In: Buck C, Demmler G, Kaplan S (eds) Textbook of pediatric infectious diseases. Elsevier, pp 2283–2304

Cocks BG, Chang CC, Carballido JM, Yssel H, de Vries JE, Aversa G (1995) A novel receptor involved in T-cell activation. Nature 376:260–263

Colf LA, Juo ZS, Garcia KC (2007) Structure of the measles virus hemagglutinin. Nat Struct Mol Biol 14:1227–1228

Condack C, Grivel JC, Devaux P, Margolis L, Cattaneo R (2007) Measles virus vaccine attenuation: suboptimal infection of lymphatic tissue and tropism alteration. J Infect Dis 196:541–549

Corey EA, Iorio RM (2007) Mutations in the stalk of the measles virus hemagglutinin protein decrease fusion but do not interfere with virus-specific interaction with the homologous fusion protein. J Virol 81:9900–9910

Crennell S, Takimoto T, Portner A, Taylor G (2000) Crystal structure of the multifunctional paramyxovirus hemagglutinin-neuraminidase. Nat Struct Biol 7:1068–1074

de Swart RL, Ludlow M, de Witte L, Yanagi Y, van Amerongen G, McQuaid S, et al (2007) Predominant infection of CD150$^+$ lymphocytes and dendritic cells during measles virus infection of macaques. PLoS Pathog 3:1771–1781

Devaux P, Buchholz CJ, Schneider U, Escoffier C, Cattaneo R, Gerlier D (1997) CD46 short consensus repeats III and IV enhance measles virus binding but impair soluble hemagglutinin binding. J Virol 71:4157–4160

Dorig RE, Marcil A, Chopra A, Richardson CD (1993) The human CD46 molecule is a receptor for measles virus (Edmonston strain). Cell 75:295–305

Erlenhoefer C, Wurzer WJ, Loffler S, Schneider-Schaulies S, ter Meulen V, Schneider-Schaulies J (2001) CD150 (SLAM) is a receptor for measles virus but is not involved in viral contact-mediated proliferation inhibition. J Virol 75:4499–4505

Fournier P, Brons NH, Berbers GA, Wiesmuller KH, Fleckenstein BT, Schneider F, et al (1997) Antibodies to a new linear site at the topographical or functional interface between the haemagglutinin and fusion proteins protect against measles encephalitis. J Gen Virol 78:1295–1302

Funke S, Maisner A, Mühlebach MD, Koehl U, Grez M, Cattaneo R, et al (2008) Targeted cell entry of lentiviral vectors. Mol Ther 16(8):1427–1436

Griffin DE (2007) Measles virus. In: Knipe DM, Howley PM (eds) Fields' virology. Lippincott Williams and Wilkins, Philadelphia, pp 1551–1585

Hallak LK, Merchan JR, Storgard CM, Loftus JC, Russell SJ (2005) Targeted measles virus vector displaying Echistatin infects endothelial cells via alpha(v)beta3 and leads to tumor regression. Cancer Res 65:5292–5300

Hammond AL, Plemper RK, Zhang J, Schneider U, Russell SJ, Cattaneo R (2001) Single-chain antibody displayed on a recombinant measles virus confers entry through the tumor-associated carcinoembryonic antigen. J Virol 75:2087–2096

Hashiguchi T, Kajikawa M, Maita N, Takeda M, Kuroki K, Sasaki K, et al (2007) Crystal structure of measles virus hemagglutinin provides insight into effective vaccines. Proc Natl Acad Sci U S A 104:19535–19540

Hsu EC, Sarangi F, Iorio C, Sidhu MS, Udem SA, Dillehay DL, et al (1998) A single amino acid change in the hemagglutinin protein of measles virus determines its ability to bind CD46 and reveals another receptor on marmoset B cells. J Virol 72:2905–2916

Hsu EC, Sabatinos S, Hoedemaeker FJ, Rose DR, Richardson CD (1999) Use of site-specific mutagenesis and monoclonal antibodies to map regions of CD46 that interact with measles virus H protein. Virology 258:314–326

Hsu EC, Iorio C, Sarangi F, Khine AA, Richardson CD (2001) CDw150(SLAM) is a receptor for a lymphotropic strain of measles virus and may account for the immunosuppressive properties of this virus. Virology 279:9–21

Hu A, Cathomen T, Cattaneo R, Norrby E (1995) Influence of N-linked oligosaccharide chains on the processing, cell surface expression and function of the measles virus fusion protein. J Gen Virol 76:705–710

Kielian M, Rey FA (2006) Virus membrane-fusion proteins: more than one way to make a hairpin. Nat Rev Microbiol 4:67–76

Lamb RA (1993) Paramyxovirus fusion: a hypothesis for changes. Virology 197:1–11

Lamb RA, Parks GD (2007) Paramyxoviridae: the viruses and their replication. In: Knipe DM, Howley PM (eds) Fields virology. Lippincott Williams and Wilkins, Philadelphia, pp 1449–1496

Lawrence MC, Borg NA, Streltsov VA, Pilling PA, Epa VC, Varghese JN, et al (2004) Structure of the haemagglutinin-neuraminidase from human parainfluenza virus type III. J Mol Biol 335:1343–1357

Lecouturier V, Fayolle J, Caballero M, Carabana J, Celma ML, Fernandez-Munoz R, et al (1996) Identification of two amino acids in the hemagglutinin glycoprotein of measles virus (MV) that govern hemadsorption HeLa cell fusion, and CD46 downregulation: phenotypic markers that differentiate vaccine and wild-type MV strains. J Virol 70:4200–4204

Leonard VHJ, Sinn PL, Hodge G, Miest T, Devaux P, Oezguen N, et al (2008) Epithelial cell receptor-blind measles virus remains virulent but cannot cross epithelia and is not shed. J Clin Invest 118(7):2448–2458

Liu TC, Galanis E, Kirn D (2007) Clinical trial results with oncolytic virotherapy: a century of promise, a decade of progress. Nat Clin Pract Oncol 4:101–117

Ludwig K, Schade B, Bottcher C, Korte T, Ohlwein N, Baljinnyam B, et al (2008) Electron cryomicroscopy reveals different F1+F2 protein states in intact parainfluenza virions. J Virol 82:3775–3781

Manchester M, Gairin JE, Patterson JB, Alvarez J, Liszewski MK, Eto DS, et al (1997) Measles virus recognizes its receptor CD46, via two distinct binding domains within SCR1–2. Virology 233:174–184

Masse N, Ainouze M, Neel B, Wild TF, Buckland R, Langedijk JP (2004) Measles virus (MV) hemagglutinin: evidence that attachment sites for MV receptors SLAM and CD46 overlap on the globular head. J Virol 78:9051–9063

McGinnes LW, Morrison TG (2006) Inhibition of receptor binding stabilizes Newcastle disease virus HN and F protein-containing complexes. J Virol 80:2894–2903

Mühlebach MD, Leonard VHJ, Cattaneo R (2008) The measles virus fusion protein transmembrane region controls formation of an active glycoprotein complex and fusion efficiency. J Virol (in press)

Nakamura T, Peng KW, Vongpunsawad S, Harvey M, Mizuguchi H, Hayakawa T, et al (2004) Antibody-targeted cell fusion. Nat Biotechnol 22:331–336

Nakamura T, Peng KW, Harvey M, Greiner S, Lorimer IA, James CD, Russell SJ (2005) Rescue and propagation of fully retargeted oncolytic measles viruses. Nat Biotechnol 23: 209–214

Naniche D, Varior-Krishnan G, Cervoni F, Wild TF, Rossi B, Rabourdin-Combe C, Gerlier D (1993) Human membrane cofactor protein (CD46) acts as a cellular receptor for measles virus. J Virol 67:6025–6032

Navaratnarajah CK, Vongpunsawad S, Oezguen N, Stehle T, Braun W, Hashiguchi T, et al (2008) Dynamic Interaction of the measles virus hemagglutinin with its receptor signaling lymphocytic activation molecule (SLAM, CD150). J Biol Chem 283:11763–11771

Ohno S, Seki F, Ono N, Yanagi Y (2003) Histidine at position 61 and its adjacent amino acid residues are critical for the ability of SLAM (CD150) to act as a cellular receptor for measles virus. J Gen Virol 84:2381–2388

Ono N, Tatsuo H, Hidaka Y, Aoki T, Minagawa H, Yanagi Y (2001a) Measles viruses on throat swabs from measles patients use signaling lymphocytic activation molecule (CDw150) but not CD46 as a cellular receptor. J Virol 75:4399–4401

Ono N, Tatsuo H, Tanaka K, Minagawa H, Yanagi Y (2001b) V domain of human SLAM (CDw150) is essential for its function as a measles virus receptor. J Virol 75:1594–1600

Panum P (1939) Observations made during the epidemic of measles on the Faroe Islands in the year 1846. Med Classics 3:803–886

Peng KW, Donovan KA, Schneider U, Cattaneo R, Lust JA, Russell SJ (2003) Oncolytic measles viruses displaying a single-chain antibody against CD38, a myeloma cell marker. Blood 101:2557–2562

Plemper RK, Hammond AL, Cattaneo R (2000) Characterization of a region of the measles virus hemagglutinin sufficient for its dimerization. J Virol 74:6485–6493

Plemper RK, Hammond AL, Cattaneo R (2001) Measles virus envelope glycoproteins hetero-oligomerize in the endoplasmic reticulum. J Biol Chem 276:44239–44246

Plemper RK, Hammond AL, Gerlier D, Fielding AK, Cattaneo R (2002) Strength of envelope protein interaction modulates cytopathicity of measles virus. J Virol 76:5051–5061

Rager M, Vongpunsawad S, Duprex WP, Cattaneo R (2002) Polyploid measles virus with hexameric genome length. EMBO J 21:2364–2372

Richardson C, Hull D, Greer P, Hasel K, Berkovich A, Englund G, et al (1986) The nucleotide sequence of the mRNA encoding the fusion protein of measles virus (Edmonston strain): a comparison of fusion proteins from several different paramyxoviruses. Virology 155:508–523

Schneider U, Bullough F, Vongpunsawad S, Russell SJ, Cattaneo R (2000) Recombinant measles viruses efficiently entering cells through targeted receptors. J Virol 74:9928–9936

Schneider-Schaulies J, Schnorr JJ, Brinckmann U, Dunster LM, Baczko K, Liebert UG, et al (1995) Receptor usage and differential downregulation of CD46 by measles virus wild-type and vaccine strains. Proc Natl Acad Sci U S A 92:3943–3947

Springfeld C, von Messling V, Frenzke M, Ungerechts G, Buchholz CJ, Cattaneo R (2006) Oncolytic efficacy and enhanced safety of measles virus activated by tumor-secreted matrix metalloproteinases. Cancer Res 66:7694–7700

Tahara M, Takeda M, Shirogane Y, Hashiguchi T, Ohno S, Yanagi Y (2008) Measles virus infects both polarized epithelial and immune cells by using distinctive receptor-binding sites on its hemagglutinin. J Virol 82:4630–4637

Tatsuo H, Ono N, Tanaka K, Yanagi Y (2000) SLAM (CDw150) is a cellular receptor for measles virus. Nature 406:893–897

Ungerechts G, Springfeld C, Frenzke ME, Lampe J, Johnston PB, Parker WB, et al (2007a) Lymphoma chemovirotherapy: CD20-targeted and convertase-armed measles virus can synergize with fludarabine. Cancer Res 67:10939–10947

Ungerechts G, Springfeld C, Frenzke ME, Lampe J, Parker WB, Sorscher EJ, Cattaneo R (2007b) An immunocompetent murine model for oncolysis with an armed and targeted measles virus. Mol Ther 15:1991–1997

von Messling V, Svitek N, Cattaneo R (2006) Receptor (SLAM [CD150]) recognition and the V protein sustain swift lymphocyte-based invasion of mucosal tissue and lymphatic organs by a morbillivirus. J Virol 80:6084–6092

Vongpunsawad S, Oezgun N, Braun W, Cattaneo R (2004) Selectively receptor-blind measles viruses: identification of residues necessary for SLAM- or CD46-induced fusion and their localization on a new hemagglutinin structural model. J Virol 78:302–313

Watanabe M, Hirano A, Stenglein S, Nelson J, Thomas G, Wong TC (1995) Engineered serine protease inhibitor prevents furin-catalyzed activation of the fusion glycoprotein and production of infectious measles virus. J Virol 69:3206–3210

Wild TF, Malvoisin E, Buckland R (1991) Measles virus: both the haemagglutinin and fusion glycoproteins are required for fusion. J Gen Virol 72:439–442

Yanagi Y, Takeda M, Ohno S (2006) Measles virus: cellular receptors, tropism and pathogenesis. J Gen Virol 87:2767–2779

Yin HS, Wen X, Paterson RG, Lamb RA, Jardetzky TS (2006) Structure of the parainfluenza virus 5 F protein in its metastable, prefusion conformation. Nature 439:38–44

Yuan P, Thompson TB, Wurzburg BA, Paterson RG, Lamb RA, Jardetzky TS (2005) Structural studies of the parainfluenza virus 5 hemagglutinin-neuraminidase tetramer in complex with its receptor, sialyllactose. Structure 13:803–815

Chapter 5
The Measles Virus Replication Cycle

B.K. Rima(✉) and W.P. Duprex

Contents

Introductory Remarks	78
Ploidy and Particle-to-Plaque-Forming Unit Ratios	78
The RNP	79
Transport of the RNP from the Plasma Membrane to the Intracellular Sites for Transcription and Replication	79
The Genome and Viral Evolution	81
Transcription	83
Virus and Host Proteins Associated with the RNP	84
General Description of the Transcription Process	86
Modifications of mRNAs	87
Phase and Transcription	89
Dynamics of RNA Accumulation and Transcription Attenuation	89
Replication	93
Translation	94
Getting It All Together	96
References	97

Abstract This review describes the two interrelated and interdependent processes of transcription and replication for measles virus. First, we concentrate on the ribonucleoprotein (RNP) complex, which contains the negative sense genomic template and in encapsidated in every virion. Second, we examine the viral proteins involved in these processes, placing particular emphasis on their structure, conserved sequence motifs, their interaction partners and the domains which mediate these associations. Transcription is discussed in terms of sequence motifs in the template, editing, co-transcriptional modifications of the mRNAs and the phase of the gene start sites within the genome. Likewise, replication is considered in terms of promoter strength, copy numbers and the remarkable plasticity of the system. The review emphasises what is not known or known only by analogy rather than by direct experimental evidence in the MV replication cycle and hence where

B.K. Rima
Centre for Infection and Immunity, School of Medicine, Dentistry and Biomedical Sciences, Queen's University Belfast, Belfast BT9 7BL, Northern Ireland, UK, e-mail: b.rima@qub.ac.uk

additional research, using reverse genetic systems, is needed to complete our understanding of the processes involved.

Introductory Remarks

The focus of this chapter centres on the ribonucleoprotein (RNP) complex, which is the basic unit of infectivity of measles virus (MV) and deals with the two intimately linked processes of transcription and replication. The preceding chapters by Y. Yanagi, C. Kemper and J.P. Atkinson, and C. Navaratnarajah et al. in this volume on membrane receptor interactions have dealt with the processes of attachment and fusion, so we start our discussion and description of these processes immediately after fusion of the virion and host cell membrane, i.e. when the RNP containing the negative sense genome has entered the cell. The RNP has a density of 1.30 g/cm^3 in CsCl gradients and is comprised for more than 97% protein (Stallcup et al. 1979), which protects the RNA from RNAses (Andzhaparidze et al. 1987). In addition, the RNP is the template for transcription and replication processes.

In this volume, the structure of the nucleocapsid core is dealt with separately (see the chapter by S. Longhi, this volume), although for the sake of completeness occasionally we refer to structural aspects to illustrate mechanistic aspects of the two processes. Accompanying chapters have focussed on host factors involved in the replication of the virus (see the chapter by D. Gerlier and H. Valentin, this volume) and the ability of the virus to combat innate immune defence mechanisms of the cells and the host organism (see the chapter by D. Gerlier and H. Valentin, this volume), and again these aspects are only touched upon tangentially in this review of the MV replication cycle.

Ploidy and Particle-to-Plaque-Forming Unit Ratios

Early electron microscopy (EM) studies showed that MV populations are pleomorphic, being comprised of spherical particles with different diameters of 300–1000 nm (Casali et al. 1981; Lund et al. 1984). Interestingly, although this was an accepted part of MV biology, it took many years to demonstrate that this heterogeneity could be accompanied by incorporation of more than one RNP in a virion. Thus, many particles contain more than one functional RNP and are therefore functionally polyploid (Rager et al. 2002). Furthermore, the particle-to-plaque-forming unit (p.f.u.) ratio of MV is high (>10:1) (Afzal et al. 2003) and hence the numbers of defective RNPs introduced in a cell upon infection is not known. These defective RNPs can either be full-length or contain large internal deletions such as are found in defective interfering particles. The latter have been demonstrated to be present in both virus stocks and vaccines, where they affect replication of the standard virus significantly (Rima et al. 1977). Virions can also contain (+)RNPs, encapsidated, antigenomic, positive-sense full-length RNA molecules. These (+)RNPs are essential intermediates in virus replication since they are required to generate progeny (–)RNPs. At present, it is thought that the budding process is unable to discriminate between (+) and (–) RNPs and genome containing RNPs are not preferentially packaged. This suggests that cells infected at a multiplicity of infection (MOI) of greater than 0.1 will contain a number of RNPs, many of which may be defective. Nothing is known about competition between multiple RNPs or whether there exists a defined, saturable set of intracellular replications site where the MV replication can progress. It remains unknown if the virus uses specific sites on the plasma membrane to enter a cell, and hence there is no information on potential coupling of entry and intracellular replication sites mediated, for example, by activities of the cytoskeleton.

The RNP

Although it is clear that more than one RNP can be found in a cell for the sake of simplicity, we will describe the ensuing processes of viral RNA synthesis, i.e. transcription and replication and mRNA translation assuming that a single RNP containing the full-length genome has entered the cell.

In addition to the genomic RNA, the RNP is comprised of the nucleocapsid (N) protein, the phospho- (P) protein and the large (L) protein (Table 1). The RNA-dependent RNA polymerase (RdRp) complex consists of the L protein and the P protein and functions both as the viral transcriptase and replicase. After negative staining and EM, the RNP appears as a helical nucleocapsid approximately 1 µm in length and 18–21 nm in diameter, with a central core of 5 nm (Waterson 1962). How the RNA and proteins are arranged in the RNP is not yet clear, but what is known is addressed in the chapter by S. Longhi, this volume. Three-dimensional reconstructions, calculated from cryo-negative stain transmission electron micrographs, indicate that the RNP exhibits extensive conformational flexibility (Bhella et al. 2004). As yet, X-ray crystallographic studies, such as those for lyssaviruses and Borna disease virus (Albertini et al. 2006; Green et al. 2006; Luo et al. 2007; Rudolph et al. 2003), have not yet been reported for MV. The MV RNP displays the characteristic herringbone structure of all paramyxovirus nucleocapsids. Electron microscopy studies of replicating intermediates have demonstrated that the herringbone points to the 5' prime end of the genome (Thorne and Dermott 1977).

Transport of the RNP from the Plasma Membrane to the Intracellular Sites for Transcription and Replication

The first question that arises is: does replication occur diffusely throughout the cell cytoplasm or within dedicated viral factories? When the N and P proteins are localised by indirect immunofluorescence (IIF) both in acutely or persistently infected cells, the resulting intra-cytoplasmic staining pattern is punctuate (Duprex et al.

Table 1 Annotation of the measles virus genome sequence

Numbers in antigenome (positive sense)	RNA	Open reading frame	Number of codons	Protein
1–52	Leader	–	–	There is a small ORF of potentially 4 amino acids.
53–55	Leader N Ig			
56–1744	N gene	108–1682 (N)	525	Nucleocapsid protein: phosphorylated protein which encapsidates the RNA and protects it from RNAses.
1745–1747	N-P Ig			
1748–3402	P/V/C gene	1807–3327 (P)	507	Phosphoprotein associated with RNP in polymerase complex as well as a chaperone for the N protein.
		1807–2702 +1G (V)	231 + 69 = 300	Prevents interferon induced transcriptional responses in MV
		1807–2702 +4G (V)	231 + 70 = 301	
		1829–2386 (C)	186	Prevents interferon-induced transcriptional responses; acts as an infectivity factor; inhibits transcription
3403–3405	P-M Ig			
3406–4872	M gene	3438–4442 (M)	335	Matrix protein: hydrophobic protein on inner leaflet of membrane; inhibits transcription
4873–4875	M-F Ig			
4876–7247	F gene	5458–7107 (F)	550	Fusion glycoprotein cleaved to a disulphide linked F2-F1 complex by furin-like proteases that are generally available; acylated
7248–7250	F-H Ig			
7251–9208	H gene	7271–9121 (H)	617	Haemagglutinin: attachment glycoprotein that interacts directly with entry receptors
9209–9211	H-L IG			
9212–15854	L gene	9234–15 782 (L)	2183	Large protein in RdRp complex with P; RNA synthesis; capping and polyadenylation
15855–15857	L-tr Ig?			
15858–15894	Trailer			Transcript not identified

Ig nontranscribed intergenic sequence in positive sense

1999; Ludlow et al. 2005). These inclusions appear to be the sites where mRNAs encoding the N-, P- and matrix (M) proteins are localised using in situ hybridisation (ISH). Furthermore, it is clear that when we studied the distribution of L proteins that were tagged with enhanced green fluorescent protein (EGFP) or with a short c-Myc epitope tag, that these are also distributed unevenly and show a similar punctuate staining patterns (Duprex et al. 2002). Indirect immunofluorescence has been used to prove that the L protein co-localises with N and P and, more surprisingly, the M protein (M. Ludlow et al., unpublished observations). All these pieces of evidence suggest that specific viral factories exist in the cell. We have also observed an association of the L protein with the plasma membrane in cells acutely infected with the closely related rinderpest virus (Brown et al. 2005) and in persistently infected cells (M. Ludlow et al., unpublished observations).

The main site for localisation of these large viral factories is perinuclear, especially in the early stages of infection, and hence the question has arisen whether these factories are associated with mitochondrial organising centres and aggresomes, as is the case in other viral infections (Wileman 2007). Currently, it is not clear if the incoming RNP is actively transported to these sites due to an interaction with components of, for example, the cytoskeletal networks. MV can grow, albeit at reduced titres, in cells treated with cytochalasin B, which leads to enucleation of cells and severe effects on the cytoskeleton (Follett et al. 1976), suggesting that if there is a role for these components they are not absolutely required. In our study of MV infection in astrocytoma cell lines using a recombinant MV which expresses EGFP (MVeGFP), no alteration of the actin, tubulin, or vimentin components of the cytoskeleton was observed in either cell type, whereas a disruption of the glial-fibrillary-acidic protein filament network was noted in MVeGFP-infected U-251 cells (Duprex et al. 2000a).

The Genome and Viral Evolution

The MV genome is a nonsegmented RNA molecule of negative polarity, i.e. opposite polarity to viral messenger RNAs (mRNAs), with a length of 15,894 nucleotides (Blumberg et al. 1988). It contains six transcription units (genes), which are separated from each other by trinucleotide intergenic (Ig) sequences, and 3' and 5' terminal sequences containing genomic and anti-genomic promoters, respectively. The gene order was originally inferred from Northern blot analysis of mono- and polycistronic mRNA (Barrett and Underwood 1985; Richardson et al. 1985; Rima et al. 1986) and thereafter by genomic sequencing (Bellini et al. 1985, 1986). The transcription units encode at least eight proteins, six of which are components of the virion (Fig. 1 and Table 1). Protruding from the envelope are glycoprotein spikes consisting of the fusion (F) protein and the haemagglutinin (H) protein (see the chapter by C. Navaratnarajah et al., this volume). The inner leaflet of the virion membrane is coated with the hydrophobic M protein. At least two additional proteins are generated from the gene encoding the P protein. The C protein is translated

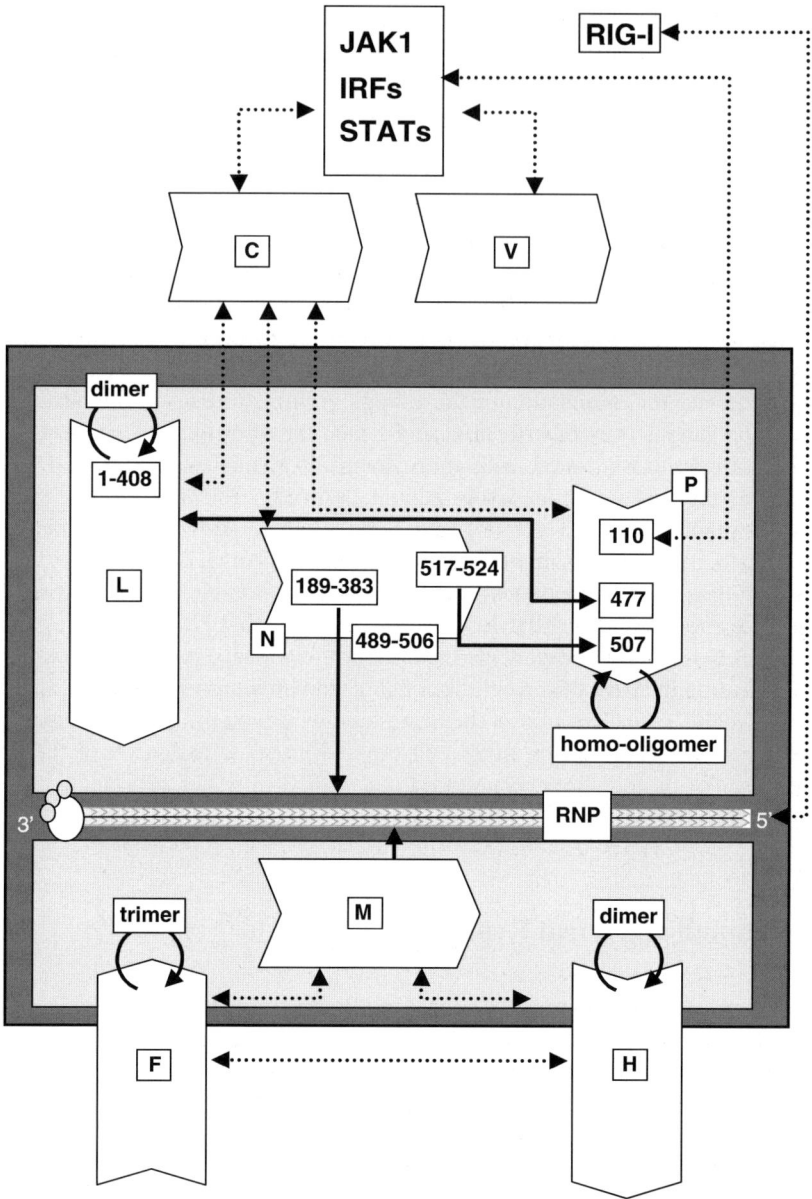

Fig. 1 The interactome of measles proteins with each other and selected host proteins. The interaction between viral proteins is depicted with a *broken arrow* if the domains have not been specified on both binding partners or when they are inferred from experimental observations, e.g. the effect of the C protein on the transcription–replication complex. A *solid arrow* indicates that the interacting domains on both partners have been determined by mutational analysis. Specific interactions of viral and host proteins are detailed in the chapter by D. Gerlier and H. Valentin, this volume

from the P mRNA from an overlapping open reading frame (ORF) (Bellini et al. 1985). The V protein is generated from an edited transcript of the P gene (Cattaneo et al. 1989a) (Table 1).

There are at least 22 known MV genotypes which are grouped in eight clades (Riddell et al. 2005; Rima et al. 1995). Seventeen genotypes are currently active in the world, and the remainder are probably extinct. So far, it has not been possible to link genotypic differences to specific biological phenotypes. There was early speculation that African B genotypes were responsible for the high levels of mortality associated with the infection on that continent, but, after more detailed epidemiological and sociocultural analysis, this has not been shown to be the case. No differences have been observed between genotypes in genome organisation or in the basic processes of transcription and replication. Furthermore, the lengths of all genes are conserved; there are no variations in editing, although this has been studied only for limited number of virus strains; the Ig sequences are conserved and, apart from subacute sclerosing panencephalitis (SSPE) strains, the transcription process is conserved. Morbilliviruses are unique within the *Paramyxoviridae* in that they contain a very long noncoding region of nearly 1 kb between the M and F ORFs (Bellini et al. 1985; Richardson et al. 1986; Table 1). The region is GC rich and is likely to fold into complex secondary RNA structures. The function of this region is not known, although the sequence is largely conserved in length and the position of the nontranscribed Ig trinucleotide spacer is also fixed since it is located almost in the middle of this region (Table 1). This long untranslated region (UTR) is a unique feature of morbilliviruses and, although dispensable for the growth of the virus in vitro (Takeda et al. 2005), the suggested functions include mRNA stabilisation and regulation of translation.

All MV strains appear to show the CpG and UpA suppression that is characteristic of most mammalian RNA viruses (Rima and McFerran 1997). In the case of MV, the CpG motifs that are most stimulatory for toll-like receptors (TLRs) are 4.3 times as infrequent as the ones that are least stimulatory. Recently, it has been shown that RNA oligonucleotides are able to stimulate the innate immune system, although the interacting partners are not well established and this may involve as yet unknown TLR-like molecules (Sugiyama et al. 2005). UpA suppression, which is also widespread throughout the RNA viruses, may be linked to the fact that in cells in which the innate immune responses are elicited, RNaseL may target RNA molecules containing this sequence for destruction (Washenberger et al. 2007).

In conclusion, it is clear that the basic processes for growth and replication of MV virus are conserved throughout all genotypes.

Transcription

In describing the processes of MV transcription and replication, it must be remembered that much of this is derived by analogy to the better studied paramyxoviruses and other members of the *Mononegavirales*, particularly vesicular stomatitis virus

(VSV) and Sendai virus (SeV) (for reviews see Lyles and Rupprecht 2006; Lamb and Parks 2006). The difficulty in reliably obtaining high MOIs, the relative low level of gene expression as well as the lack of a dependable in vitro transcription system has hindered the study of molecular biology of MV.

In the genome, the six transcription units are preceded by a region at the 3'-terminus coding for the 56 nucleotide leader RNA and are followed at the 5'-terminus by a trailer region of 40 nucleotides (Blumberg et al. 1988). The sizes of the various transcription units and the coding regions, and the derived sizes of the various proteins, are given in Table 1. The entire nucleotide sequence has been compiled for the measles prototype strain Edmonston (Radecke and Billeter 1995), which was isolated in 1954 and passaged in tissue culture by John Enders (Enders and Peebles 1954). In total approximately 30 vaccine and wild-type strains have been completely sequenced.

Virus and Host Proteins Associated with the RNP

The RNP consists of the genome or antigenome RNA and 2,649 copies of the N protein. This figure is based on the fact that one protein can protect six nucleotides (Calain and Roux 1993). Associated with this structure are two additional viral proteins: the P protein, in approximately one-tenth that number, and roughly 20–30 copies of the L protein (Lamb et al. 1976; Portner et al. 1988). Every virion must contain a nucleocapsid/L/P complex to ensure infectivity. The L protein contains the enzymatic activities of the RdRp, which acts as both the viral transcriptase and replicase. At least two additional nonstructural proteins are generated from the gene encoding the P protein. The C protein is translated from the P mRNA from an overlapping ORF (Bellini et al. 1985). The initiation codons for P and C are not in optimal context, and hence cap-dependent scanning appears to allow the ribosome to initiate translation at either the start of the P open reading frame at position 60 in the mRNA or that of the C ORF at position 82. The C protein is composed of 186 amino acids and it has been proposed to function both as an infectivity factor and an inhibitor of type I interferon-induced transcriptional responses and signalling (see the chapter by D. Gerlier and H. Valentin, this volume). Virus mutants lacking this protein can be propagated successfully in tissue culture (Radecke and Billeter 1996). However, more relevant to the present discussion, the C protein has been shown to inhibit expression of reporter genes in minigenome rescue assays (Bankamp et al. 2005). As these are combined replication and transcription assays, the actual step that is inhibited by the C protein of MV is not clear.

The V protein is generated from an edited transcript of the P gene (Cattaneo et al. 1989a). A recombinant virus which fails to express this protein has been generated (Schneider et al. 1997). Interference by with STAT-inducible transcription, which may lead to virus-induced cytokine inhibition in vivo, and promotion of STAT1 degradation have been observed (see the chapter by D. Gerlier and

H. Valentin, this volume). It is amino-co-terminal with the P protein for the first 231 amino acids and therefore shares the main phosphorylation sites of P, including the important tyrosine residue at position 110 (Devaux et al. 2007; Fontana et al. 2008). The role of tyrosine 110 in modulating interactions with the innate immune system is discussed in detail in the chapter by D. Gerlier and H. Valentin, this volume. A 69-residue cysteine-rich tail, which contains a zinc-binding finger domain, follows the common amino terminal 231 residues (Liston and Briedis 1994). This domain represents one of the most conserved products encoded by the P genes of paramyxoviruses. The V protein also has been implicated in affecting the expression of reporter genes in the minigenome expression assay (Parks et al. 2006; Witko et al. 2006), but this has not been confirmed in other studies. Thus, though the V and C proteins may affect the gene expression in these assays, viruses in which the expression of these proteins is ablated grow well in tissue culture, indicating that they are not essential components of the RdRp.

The RdRp can only use the RNP and not naked RNA as a template (Rozenblatt et al. 1979). Although the L/P complex contains the requisite catalytic activities for the RdRp, its activity may also involve or be modulated by host cell proteins and/or cytoskeletal components such as tubulin, which has been reported to stimulate in vitro RNA synthesis (Horikami and Moyer 1991). The stoichiometry of the complex is not known for MV and has been suggested to be L/P_4 for SeV and L/P_2 for VSV (Chen et al. 2006; Curran and Kolakofsky 2008). The P protein of the closely related morbillivirus RPV is a tetramer (Rahaman et al. 2004). Oligomerisation of the MV L protein has been inferred (Cevik et al. 2003), but this remains to be demonstrated directly. The interaction of the RdRp complex with the RNP probably involves a bridging function for the P protein, which has binding sites for both the N and L proteins. Direct L–N interactions have not been documented. In Fig. 1, we have attempted to synthesise the interactome of MV proteins with themselves (homo-oligomerisation), each other and with selected host factors (hetero-oligomerisation). The complexity of the diagram indicates the plethora of interactions exhibited by viral proteins even in a virus, which only expresses eight proteins. Apart from the N and P and L interactions, relatively few direct deletion analyses have been performed to locate the interacting sequences. The interactions of M with the F and H proteins and the oligomerisation of the glycoproteins themselves are described in the chapter by C. Navaratnarajah et al., this volume.

The original comparison by Poch et al. (1990) of five L proteins from *Mononegavirales* identified six (I–VI) conserved domains. Few have been linked to specific functions apart from domain III, which contains the canonical sequence GDNQ as a potential RdRp active site. For the morbilliviruses, we refined this analysis and found that there are two nonconserved hinges in the L protein separating three largely conserved domains (McIlhatton et al. 1997). The first (D1) contains the Poch domains I and II, the second (D2) domains III, IV and V, and the third (D3) containing conserved domain VI. The latter has recently been shown to be part of a structure involved in 2-O-methyl transferase activity (Ferron et al. 2002). Sequence comparison of the L proteins of vaccine and wild-type strains of MV confirmed that genotypic differences exist but no functional differences were observed, and this study was not able to identify consistent changes between vaccine and wild-type strains (Bankamp et al. 1999).

In the *Mononegavirales* prototype, VSV, there is evidence that the transcriptase complex has a number of host proteins associated with it such as EF1α (Qanungo et al. 2004), which is not present in the replicase version of the complex. However, to date this has not been confirmed for paramyxoviruses, and the question of whether the transcriptase complex is the same as the replicase complex has not been resolved. Furthermore, although two forms of the RdRp complex may contain the same proteins, it is possible that they might differ in the phosphorylation state. Whilst the L protein of MV is not phosphorylated, both the N and P proteins are. Phosphorylation of the N protein, and whether this plays any role in the viral life cycle, has not been studied. It is known that threonine and serine but not tyrosine residues are involved. The same is the case for the P protein, where there are multiple phosphorylation sites. At least one of serine residues at positions 86, 151 and 180 are phosphorylated by casein kinase II. Tyrosine 110 is also phosphorylated (Devaux et al. 2007). Temporal analyses to gain insight in potential controlling roles for phosphorylation and function studies of the phosphorylation of viral proteins using reverse genetics remain to be done.

General Description of the Transcription Process

In the genome, the six transcription units are arranged in the order 3′-N-P(V/C)-M-F-H-L-5′ (Dowling et al. 1986; Rima et al. 1986). The sequence of the nontranscribed intergenic regions between the first five genes in the genome (negative strand) is 3′-GAA-5′. There is a single nucleotide difference in the sequence between the H and L genes which is 3′-GCA-5′ (negative sense). These Ig trinucleotides are not transcribed by the RdRp. The genome consists of a multiple of six nucleotides (6 × 2,649). A suggested explanation for the rule of six (Calain and Roux 1993), which is obeyed by MV (Sidhu et al. 1995), is that each N protein associates with exactly six nucleotides of RNA. The 2649 nucleocapsid proteins and the RNA form a helical RNP with 2649/13 = ~204 helical turns. The number of helical turns was determined to be 204 (Lund et al. 1984), remarkably accurately and long before the rule of six and the number of N proteins per helical turn were established.

In MV there two recognised sequence elements for the promoter sequence, the so-called A box which comprises the first 15 nucleotides at the 3′ end of the genome and antigenome, and the B box, which is represented by the sequence $GN_5GN_5GN_5$ at positions in the 14, 15 and 16th hexamers (see the chapter by S. Longhi, this volume). Structural studies of the RNP suggest this would place the B box immediately beside the A box in the first and second helical turn of the RNP, which has been proposed to contain approximately 13 N protein molecules per turn and hence 13 × 6 = 78 nucleotides (Lamb and Parks 2006). Mutagenesis has demonstrated that the three G residues in the B box are essential and this element is conserved in all morbilliviruses (Rennick et al. 2007; Walpita 2004). Liu et al. (2006) analysed the genome termini of a large number of MV vaccine and wild-type strains and changes were only found in the 3' genome terminus at positions 26 and 42. Minigenome expression assays indicate that these naturally occurring nucleotide variations in the 3' leader region affected the levels of reporter protein synthesis (Liu et al. 2006).

It is not clear whether transcription always starts at the 3' end of the template producing the 56 nucleotide leader or if a capped mRNA can be generated directly from the start of the N gene. The 56-nucleotide leader RNA has been found linked to the mRNA encoding the MV N protein (Castaneda and Wong 1990). This in itself should render the leader-N transcript untranslatable, as it would place an AUG codon and an ORF encoding four amino acids upstream of the major ORF encoding the N protein. Free leader RNA has been not been demonstrated, certainly not at the high levels one may expect in infected cells if the transcription process always involved the start of copying at the 3' end of the genome, which should lead to at least equimolar amounts of the leader and N mRNA in the cells. However, there are caveats. Absence of evidence is not evidence of absence and the relative stability of such a small RNA molecule which has no 5' cap or 3' poly-A tail may make it prone to rapid degradation. Castaneda and Wong showed that there are a large number of leader-N and leader-N-P read-through RNA molecules which are polyadenylated but not capped in both acute and persistently infected cells. These are found in structures with the density of RNPs and not on polysomes (Castaneda and Wong 1990). This, and the fact that these molecules were not made when cells were treated with cycloheximide, indicates these leader-N-read-through products are probably derived from abortive replication processes rather than from transcription. They could not detect free leader and estimated their lower detection level to be 1/400th of the level expected if leader and N were expressed in equimolar amounts. Hence these data suggest that transcription starts not with a leader but with the beginning of the N gene. Recently proposed mechanisms for nonsegmented negative-strand RNA synthesis indeed suggest that the polymerase of well-studied *Mononegavirales* may be able to approach the 3' end of the template and scan until it starts to copy the template at the N start site and that the synthesis of free leader is a relatively infrequent event as is found to be the case for MV (Curran and Kolakofsky 2008).

Modifications of mRNAs

Transcription starts, or restarts in the case of the second and subsequent transcription units, with a consensus gene start (GS) sequence of AGGRNNc/aARGa/t at the 5' end of each transcript. Similar to what is observed in all *Mononegavirales*, the MV mRNAs have been shown to be capped (Hall and ter Meulen 1977). An O-methyl transferase motif has been identified in the D3 domain of the L protein of MV using bioinformatics-based approaches (Ferron et al. 2002). It is assumed that this would be involved in the formation of the methylated cap structure, which contains an inverted guanosine residue. However, direct evidence from reverse genetics has not been published to support this. As MV is a strictly cytoplasmic virus with no known involvement of the nucleus, demonstrated by the fact that the virus grows equally well in enucleated cells (Follett et al. 1976), it is assumed that splicing of mRNAs does not take place. There is indeed no evidence for this from any cloning and sequencing studies that have been undertaken on mRNAs. However, there is evidence that the N and C proteins are translocated to the nucleus, although no associated functionality has been described (Huber et al. 1991; Nishie et al. 2007). Accumulation of RNP is particularly prominent in the nuclei in infected cells in cases of SSPE (Chui et al. 1986). All mRNAs are polyadenylated with the length of the tails similar to those of normal cellular mRNAs (Hall and ter Meulen 1977). The process used to generate the poly-A tail requires a signal sequence motif at each gene end (GE). When the RdRp complex encounters the GE sequence of 5'-RUUAUAAAACTT-3' (positive sense), stutters or slips on the U-rich template sequence adding between 70–140-A residues to the 3' end of the nascent mRNA (Hall and ter Meulen 1977). The GE signal is characterised by a conserved RUU motif in all morbilliviruses followed by an A-rich block of eight nucleotides interspersed with one or two pyrimidines. No mutational analysis has been performed to assess the functional importance of any of these sequence signals in MV. Mechanistically there is no information on the process by which the RdRp stops polyadenylation when the length of the tail reaches the desired length. Interestingly, although generated by entirely different mechanisms, this size range mirrors that of cellular mRNAs.

A similar signal to the GE is found in the middle of the P gene 5'AUUAAAAAGGG-CAC-AGA3'. In this case, this leads not to polyadenylation but to the nontemplated insertion of extra G residues due to RdRp slippage or stuttering on the UUUUCCC template. This is referred to as co-transcriptional editing and leads to the generation of transcripts encoding the V and W proteins (Cattaneo et al. 1989a). Stuttering occurs to a very limited extent at the so-called editing site. The V protein is translated from an mRNA with one extra G, which shifts the ORF after the first 231 codons for P and gives rise to the unique carboxy terminal 69 amino acid extension. The protein (W) which would result from incorporation of two extra Gs has not been identified in morbillivirus-infected cells, possibly because editing is tightly controlled, meaning that in most cases editing leads to the insertion of only a single extra single G residue. In other members of the *Paramyxovirinae*, editing is less tightly controlled and a number of extra G residues (1–4) are incorporated during the process, giving rise to V-, W- and P-like proteins with extra glycine residues (Lamb and Parks 2006). The percentage of edited (V) as opposed to nonedited (P) mRNAs in acute MV-infected cells varies from 30% to 50%. This may depend on the cell type in which infection takes places. There is no evidence that other mRNAs of MV are similarly edited. The sequence motif of RYY followed by an A-rich stretch is found in several places in the genome, and we have analysed one of the at the 3' end of the H mRNA, which through editing could access conserved ORFs in all MV strains, giving rise to different C terminal extensions of the H protein. However, at the RNA level no evidence was found that indicated that the process took place at this site (position 8943).

Phase and Transcription

A consequence of the fact that precisely six nucleotides associate with a single N protein is that each nucleotide has a specific phase (positions 1, 2, 3, 4, 5 or 6) with respect to the N protein and 3' end of the genome. Minigenomes that do not obey the rule of six are neither replicated efficiently nor readily encapsidated (Sidhu et al. 1995). Comparative analysis of morbillivirus genomes has established that the phase of the transcription starts sites is conserved between the various viruses, although not between the genes (Iseni et al. 2002; Rima et al. 2005). Thus, the transcription start site for all N mRNAs of morbilliviruses is position 2, meaning that the U that resides in the template is the second one protected by the N protein molecule. The phase for the start sites of the P and L genes is also 2; for M it is 4 and for the H it is gene 3. The only variation appears to be in the phase of the start site for the F gene, as it is 3 in MV and 2 for rinderpest virus canine distemper virus (CDV), phocine distemper virus and the cetacean morbilliviruses. This raises the question: is phase important? Phasing is conserved but obviously not between genes in one genome. However, variations in morbillivirus genome lengths always involve the deletion or insertion of multiples of six nucleotides primarily, but not exclusively, within the UTRs. As such, this preserves the phasing of the genes. Hence, it appears there is a significant biological pressure for phase conservation. Furthermore, Iseni and coworkers demonstrated that phase affected co-transcriptional editing patterns in SeV (Iseni et al. 2002). This indicates that the RdRp complex has at least a potential to sense phase and encapsidation imposes a higher order structure on the genetic entity that is the RNP, similar to, for example, the histone code (Iizuka and Smith 2003).

The topology of the transcription, and for that matter the replication processes, is unknown. From first principles, the energy needed to have both the RdRp and the de novo RNA wind around the RNP as the RdRp copied the template would be significant. Topologically, this could be achieved with the least energy expenditure if the RdRp remained in a fixed position and helical RNP rotated around the axis of its length. Current models suggest that the interactions between the RNA and the N protein are either broken or that at least the RNA template is available for base recognition.

Dynamics of RNA Accumulation and Transcription Attenuation

When the RdRp reaches the GE, one of four processes may occur, the complex may:

1. Polyadenylate the mRNA, skip transcription of the intergenic sequence, commence transcription of the next gene, add an inverted guanosine residue to the 5' end of the mRNA and finally methylate the newly formed cap;
2. Polyadenylate the mRNA and drop off the template;

3. Ignore the GE, Ig and GS sequences altogether, transcribe them all and thereby allow the formation of a read-through mRNA, which can contain as many as five ORFs (5'-N-P-M-F-H-3' transcripts have been detected);
4. Start limited polyadenylation and then continue with the formation of a read-through mRNA. No evidence has been found for this latter process to occur in MV though it has been demonstrated to occur very occasionally in other *Paramyxovirinae* (Paterson et al. 1984).

The finite chance for the polymerase complex to leave the template gives rise to a gradient of gene expression in which the promoter proximal (e.g. N and P/V/C) genes are transcribed much more frequently into mRNA than the promoter distal H and especially the L gene.

Early studies examining RNA synthesis used ^{32}P orthophosphate labelling in the presence of actinomycin D to inhibit cellular but not viral RNA synthesis. Long labelling periods were required to achieve detectable amounts of incorporation, and hence it was not possible to carry our pulse chase experiments to assess rates of synthesis and degradation (Barrett and Underwood 1985). In the most successful experiments, 3- to 4-h labelling with carrier-free orthophosphate indicated that each of the viral mRNAs accumulated to levels that were in accordance with what would be expected from the transcription gradient. In one study in which cells were shifted to 39°C, at which little or no transcription takes place, it was clear from these and other experiments (Schneider-Schaulies et al. 1991) that none of the viral mRNAs decayed more quickly than the others and these experiments also showed a very low rate of turnover of the various viral RNA species in the infected cell (Ogura et al. 1987). The levels of the various mRNA species were analysed by Northern blots. Such approaches measure steady state levels at the time of harvesting the cells rather than dynamics.

Seminal studies were conducted by Cattaneo et al. (1987) and Schneider-Schaulies et al. (1989). Both used calibrated amounts of in vitro transcripts as internal controls but the two studies came to different conclusions about the steepness of the transcription gradient. The relative frequency of each of these events is approximate 0.7 for stop and restart, 0.2 for RdRp to leave the template and 0.05–0.1 for read-through in several of the *Mononegavirales*, where this process has been studied in detail, e.g. in VSV and Sendai virus (Gupta and Kingsbury 1985; Banerjee et al. 1991; Whelan et al. 2004). The paper by Cattaneo et al. indicates that in acutely infected Vero cells there are approximately 26,800 copies of the N mRNA but only approximately 9,100 copies of the P mRNA. After the P gene, the transcription gradient becomes less steep and the remainder of the mRNAs are present at amounts approximately 0.8 times the previous, e.g. 6,700 M mRNA copies, 5,400 F mRNA copies and 4,000 H mRNA copies. There appears to be a very high level of transcription attenuation at the H-L boundary, as only 400 copies of the L mRNA were detected. The Ig sequence between the H and L genes (3'-GCA-5' in the template) is different from the consensus 3'-GAA-5', and this has been suggested to be associated with very high levels of transcription termination. Alternatively, these results may reflect difficulties in the isolation of the very large L mRNA molecules during polyA selection of mRNAs or double-stranded RNA formation due to the presence of defective interfering particles which may contain parts of the L gene during the RNA isolation. In the studies by Schneider-Schaulies et al., the copy numbers of the L mRNAs were not assessed (Schneider-Schaulies et al. 1989). In general, the transcription gradient was found to be less steep and the frequency of stop and restart events was approximately 0.75 ± 0.04 at the intergenic boundaries. Plumet et al. approached the enumeration of copy numbers of viral RNAs using a carefully calibrated real-time, reverse transcription-polymerase chain reaction (RT-PCR) using primer sets that gave as near equimolar products for each of the genes as was possible (Plumet et al. 2005). Their data support those from Cattaneo et al. and show a very severe attenuation after the N gene. Infected cells were found to contain 31,000 copies of N and only 4,800 copies of the P/V mRNA.

Re-analysing the older literature – ^{32}P orthophosphate-labelled mRNAs, Northern blots and protein synthesis experiments using pulse labelling with various labelled amino-acids including ^{35}S and ^{14}C amino acids – does not indicate that there is a large difference in the amount of the N and P mRNAs. However, these experimental approaches can obviously not prove this point. One would have to assume very large variations in rates of translation, specific activities of probes for different genes, etc., to reconcile the data with the presence of very significant attenuation after the N gene. Apart from the fact that this does not make much sense in terms of protein requirements, my own attempts (B.K. Rima, unpublished data) at quantifying the mRNAs in Northern blots using the same protocol as Cattaneo et al. did not support attenuation at this point in the genome to be larger than that observed at other intergenic boundaries. In contract, in several experiments we failed to observe any attenuation as the copy number of P mRNAs was the same as that of the N mRNAs. This is similar to the result obtained by Rennick et al. who used bi-cistronic minigenomes expressing fluorescent reporter proteins (Rennick et al. 2007). Thus the precise levels of mRNA and attenuation levels remain to be determined.

Plumet et al. described four phases in the dynamics of RNA accumulation (Plumet et al. 2005). In the first phase (0–6 h.p.i.) there is linear accumulation of mRNAs, probably resulting from RdRp molecules that were attached to the template in the incoming RNP. During this phase, transcription is not sensitive to inhibition by cycloheximide. From the accumulation of various mRNAs, the authors calculated an in vivo RNA synthesis rate of three nucleotides per second (nt/s) for the MV RdRp. This corresponds well with the rates for in vitro RNA synthesis determined for VSV and Sendai virus of 3.5 and 1.7 nt/s, respectively (Barr et al. 2002; Gubbay et al. 2001). In an MV strain in which the ORF of EGFP was inserted into the L protein (Edtag-L-MMEGFPM) the calculated rate of RNA synthesis was substantially reduced to 0.8 nt/s. This precipitates a problem since the current model assumes that initiation of RNA synthesis only occurs at the 3′ end of the template RNP. If this is the case at that rate of synthesis, the H and L mRNAs could only be completed after 3 and 7 h, respectively. The authors suggest that incoming RdRp already attached to the RNP template would be able to start (or continue) transcription starting at the location in the RNP where they had stalled. However, the estimate may be incorrect as the growth curve of this virus indicated a delay of 4 h, which is not as much as would be expected from this calculation. Also, the activity in minigenome expression assays of the L protein of Edtag-L-MM_{EGFP}M was between 40% and 60% (Duprex et al. 2002). In the second phase (6–12 h.p.i.), exponential accumulation of mRNAs takes place presumably as a result of the formation of new RdRp complexes. These start to contribute to replication in the third phase (12–24 h.p.i.). In the fourth phase (24–28 h.p.i.) the RNA synthesis slows down either due to an unknown mechanism or simply as a result of the widespread fusion, syncytium formation and associated cytopathology. There is no further accumulation of viral RNA in the fifth phase (>28 h.p.i.).

The slope of the transcription gradient varies from one cell type to another and appears especially steep (little L expression) in the human central nervous system (Cattaneo et al. 1987; Schneider-Schaulies et al. 1989). For example, in some neuronal cell lines and SSPE cases, which is a rare and invariably fatal late sequela of MV infection, the slope of this gradient of gene expression is so steep that very little F and H protein are produced (Cattaneo et al. 1987). This may aid the virus in escaping detection by the immune system by disposing the virus to a wholly intracellular existence. Consequently, the amounts of L protein synthesised should also be substantially reduced in these cells. Similar observations on the enhanced steepness of the gradient of gene expression have been made in in vitro infections of neuronal cells (Schneider-Schaulies et al. 1989). These cells also respond differently from others to the expression of the interferon inducible MxA protein, which is rapidly induced in these cells to high levels. After stable transfection with MxA into human glioblastoma cells (U-87-MxA), they released 50- to 100-fold less infectious virus and expression of viral proteins was highly restricted. The overall MV-specific transcription levels were reduced by up to 90% due to an inhibition of viral RNA synthesis and a steeper gradient of gene expression and not to decreased stability. However, this effect is cell-type-specific because in other cells MxA affects translation (Schnorr et al. 1993).

An additional transcription variation in SSPE has been observed in a number of cases where the transcriptase leads to frequent or exclusive generation of a P-M read-through transcript (Baczko et al. 1986; Yoshikawa et al. 1990). In the well-studied cases, this was found not to be due to mutations in the GE, Ig or GS sequences between the P and the M gene. Hence, other effects of mutations in the N or L proteins were invoked, but not analysed, to explain these data. To date, none of this has been subjected to analysis by reverse genetics analysis. To the best of our knowledge, the generation of such a read-through transcript means that no or little M protein will be expressed as it is in the second ORF and thereby probably unavailable for translation. Reduced expression of the M protein is a general feature of SSPE and many MV virus SSPE strains appear to have undergone hypermutation in this gene, testifying to the nonessential nature of the M protein in this CNS infection (Schneider-Schaulies et al. 1995). This may be related to the fact that M protein, though required for budding, has a general negative effect on RNA synthesis in several *Mononegavirales*, including MV (Suryanarayana et al. 1994).

In conclusion, there are still questions remaining about MV transcription. The lack of a reliable in vitro system has hampered studies, although more studies on transcription using real-time RT-PCR would be helpful.

Replication

During replication, the RdRp binds at the 3′ end of the genome and the nascent RNA molecule is immediately encapsidated by N protein. On the basis of the N protein requirement for replication, the level of this protein as a free N–P complex has been put forward as an important parameter which controls the relative levels of transcription and replication (Banerjee 1987; Kingsbury 1974). Signals in the GS, GE, the Ig trinucleotide and the editing site are ignored. This gives rise to full-length positive-stranded RNA molecules within an RNP, with Y forms of the RNP being observed as intermediates by EM (Thorne and Dermott 1977). This allowed identification of the 3′ and the 5′ ends of the herringbone as these images permit the template and the product RNP to be readily discriminated. The herringbone points to the 5′ end. The positive antigenome-containing RNPs have a strong promoter at their 3′ end, which allows for a generation of excess amounts of RNPs containing the genomic RNA. The ratio between antigenome and genome containing RNPs in infected cells was estimated to be 0.43 (300/700) by Cattaneo et al. (1989b) and 0.13 (130/1030) by Plumet et al. (2005), who also showed that the same ratio occurred in virions as in infected cells and thus that there was no selection for packaging in to virions between antigenome and genome containing RNPs. Udem and Cook (1984) estimated this ratio to be 0.40, e.g. 2.5 times more genome than antigenome RNPs (Udem and Cook 1984).

The genome, antigenome, leader and trailer RNAs are the only species in the replication process that have 5′ triphosphates at their end and hence are likely to activate the innate immune system through RIG-I (Plumet et al. 2007). Molecules with 5′ triphosphates are especially enriched in virus stocks that contain large numbers of defective interfering particles. These have been demonstrated to be present in vaccine stocks and to play a role in determining the levels of interferon induction.

The A and B boxes are all that is required for the MV replication process and knowledge of that has allowed the genome to be segmented (Takeda et al. 2006). Takeda et al. split the genome into three fragments: one encoding the N and P protein plus Lac-Z, a second one the M and F genes and DsRed and the third fragment the H and L genes and EGFP. Together, the three fragments were able to replicate as a segmented MV virus. Cells infected with only two of the three segments were observed at frequencies that were so low as to indicate that most MV particles contained at least one of each of the three (and more) segments, confirming the polyploid nature of MV particles. The system allowed the expression of up to six reporter genes and illustrates the stability and apparent unlimited packaging capacity of the particles, which make this system an excellent one for the concerted delivery of multiple genes and proteins. It has also been possible to produce a single nonsegmented recombinant MV virus expressing three reporter genes (Lac-Z, EGFP and CAT) from separate additional transcription units (Zuniga et al. 2007). The fidelity of genome replication is somewhat an enigma. Mutations rates in vitro are estimated to be 9×10^{-5} (Schrag et al. 1999). However, in vivo, i.e. circulating virus, the population appears to be much more stable with a mutation rate in the N gene of 5×10^{-4} per nucleotide per year (Rima et al. 1997). The genome is very stable and ORFs present within additional transcription units are very stably maintained (Duprex et al. 1999), even during passage in vivo (de Swart et al. 2007; Duprex et al. 2000b; Ludlow et al. 2008).

Translation

Surprisingly little is known about translational control in the RNA viruses, even though this is one of only two processes which offer opportunities for these viruses to fine-tune the expression of their proteins. Global effects on translation of host mRNAs have not been reported in MV-infected cells and there is no documented effect on the phosphorylation of eIF2α, which is a prominent mechanism for other viruses to suppress host mRNA translation (Williams 1999). Additionally, no activation of PKR has been demonstrated. Studies using the MxA protein, the type I interferon-inducible human protein, which confers resistance to VSV and influenza A virus in MxA-transfected mouse 3T3 cells, showed a 100-fold reduction of released infectious virus for MV from U937 clones that constitutively express MxA. In this cell line, but not in others MV rates of transcription or the levels of MV-specific mRNAs were not affected. However, a significant reduction in the synthesis of MV glycoproteins F- and H- but not other viral proteins was observed in U937/MxA cells, indicating differential translational effects on some specific viral mRNAs induced by MxA expression.

The absence of global effects of MV on host translation is not surprising as MV is very prone to establishing noncytolytic, persistent infections which allow for a balanced synthesis of viral and host mRNAs and their translation. However, in acutely infected Vero cells, suppression of host protein synthesis is observed (Rima and Martin 1979), although this may reflect competition for ribosomes as the detailed analysis of copy numbers of MV RNA indicates that roughly 52,000 copies of MV RNA would account for approximately 25% of the total RNA population in a cell.

MV has a demonstrated potential for controlling gene expression through translational controls. The initiation codons for the P and C ORFs are in suboptimal contexts in all morbilliviruses and hence cap-dependent scanning of the P/C or V/C mRNAs appears to allow the ribosome to initiate translation at either the start of the P/V ORF at position 60 in the mRNA or that of the C ORF at position 82. In principle, this offers an opportunity to modulate the ration of C proteins over the P and V proteins. The C protein has 186 amino acids, but its length and more importantly its sequence differs in various morbilliviruses. It is not clear from the studies conducted so far if knocking out C expression by altering the initiation codon has an effect on P or V expression. Since the initiation codon for C is the second one, it is not clear whether the fact that the scanning ribosomes do not encounter the C initiation codon leads to formation of amino-terminally truncated P and V proteins or the expression of the 57 amino acid ORF starting at position 102 in the third frame, which contains a second methionine codon with a strong Kozak consensus sequence at position 165. Surprisingly altering the context of the start codon for the P ORF to a strong context from CCGAUGG to GAGAUGG did not alter the levels of P and C protein expression when the P/C gene was expressed in adenovirus vectors (Alkhatib et al. 1988). The authors interpreted this as indicative that the levels of C and P expression are not controlled by the leaky scanning of the ribosome.

One area that has been studied in relation to translational control in MV is the extended 3' UTR of the M mRNA and the 5' UTR of the F mRNA together forming a 1-kb-long unique untranslated region which is well conserved in morbillivirus genomes. Relatively little is known about the role of the 430-nt-long 3' UTR of the M mRNA. Takeda et al. have shown that it promotes M protein synthesis and thereby stimulates replication, as M protein has been shown to be an inhibitor of transcription in most *Paramyxoviridae*, including MV (Takeda et al. 2005). The effect on translation of deleting this or mRNA stability has not been reported. In our study of gene expression in minigenomes that express two autofluorescent reporters, deletion of this sequence had no effect on gene expression (Rennick et al. 2007). The function of this 3' UTR remains to be determined.

More work has been done on the 5' UTR of the F mRNA. In CDV, the region appears to encode an extremely long signal sequence, but the details of its function are not known (von Messling and Cattaneo 2002). It would be expected that this long signal sequence (if that is what it is) must be co-translationally removed as no large precursors for the F protein have been identified in pulse-chase labelling experiments (Campbell et al. 1980). In MV, the initiation codon appears to be at position 572. Contradictory results have been published on the effect of the 5' UTR. Removal of the this region from F mRNA transcripts leads to increased translation in vitro (Richardson et al. 1986), suggesting that the F-5' UTR sequence may have a negative effect on translation of the mRNA. The region has an unusually high content of cytosine residues (44%, including a $G_7C_7A_7$ tract) with an overall GC content of 64% (de Carvalho et al. 2002) and a high degree of predicted secondary structure (Curran et al. 1986) and conservation between strains. It has been suggested that the sequence acts as a focusing factor, directing translation initiation to the second of three clustered in frame AUG codons (Cathomen et al. 1995). However, mRNA transcripts with a high potential to form stable secondary structure in the 5' UTR tend to be translated inefficiently (Kozak 1991a, 1991b). It has also been shown that all known mRNA transcripts encoding ribosomal proteins have a short (5-14 nucleotide) oligopyrimidine tract at their 5' end which has been associated with their underutilisation in translation (Pain 1996). The F-5' UTR sequence contains multiple C_5 tracts towards the 5' end, which could lead to a negative effect on translation of the F mRNA transcripts. A study of the phenotype of the recombinant MV (del5F), which contains a 504-nt deletion in the F-5′ UTR in the SCID-human thymus and liver mouse model (Auwaerter et al. 1996), found that this deletion resulted in decreased peak virus production and a small change in the kinetics of growth (Valsamakis et al. 1998). The virus had an early growth advantage, but the infection led to increased thymocyte death. The authors speculated that high levels of F synthesis might lead to increased virus production early in infection but might also cause premature fusion and death of infected cells prior to peak virus production. Their conclusion was that the F-5′ UTR sequence was not absolutely required for MV replication in that model system, but that its deletion led to a reduction, but not the abrogation of F protein expression. These conclusions reflect observations using MV minigenomes expressing two fluorescent reporter genes where the presence of the F-5′ UTR led to decreased gene expression of the second reporter (Rennick et al. 2007). This effect was interpreted as indicating that the 5′ UTR of the F gene had a negative effect on translation of the gene. Interestingly a long 5′ UTR in the HN gene of human parainfluenza type I virus controls transcription read-through at the M-F gene border rather than translation (Bousse et al. 2002).

The translation of the M and F mRNAs is sensitive to elevated temperatures to a larger extent than that of the others, as its is abrogated immediately after a shift up to 39°C whilst the translation of the N, P and H protein mRNAs still occurs at this temperature, though at lower rates (Ogura et al. 1987). This effect is reversible because shift down immediately restores translation. The F mRNA also appears to be distributed on smaller polysomes than that of the H mRNA at the higher temperatures (Ogura et al. 1988), although it remains to be seen whether these effects are due to the presence of these extended UTRs.

The precise intracytoplasmic site of viral protein synthesis is essentially unknown. We have some preliminary evidence that the N, P and M mRNAs accumulate in the viral factories that have been identified earlier. Whether they are translated there or are more diffuse in the cytoplasm cannot be evaluated. It is also not clear whether transcription and translation are coupled. No information is available for the other mRNAs since their abundance is too low or too diffuse for detection by ISH. It is a reasonable assumption that the H and F mRNAs are localised at the rough endoplasmic reticulum and that these glycoproteins are translated there before processing through a pathway that involves the Golgi apparatus and vesicular transport to the cell membrane (see the chapter by C.K. Navaratnarajah et al., this volume).

Getting It All Together

Transport of the various viral components to the cell membrane and specific structures in these are processes that are not well understood, and this is exemplified best by considering what is known in SSPE. In this disease, MV is able to spread along neuronal anatomical pathways, which probably requires trans-synaptic transfer of the virus. This poses a remarkable problem for the virus. It is likely that replication and transcription takes place in the cell bodies where most of the required host cell functions are localised. The RNP would then have to be transported to the end of the neuronal axons and or dendrites. Then, for budding to take place, somehow the two viral glycoproteins also need to arrive at the same location. Location and localisation are also important in cytopathology, as our study on persistence of MV in NT2 cells indicates (Ludlow et al. 2005). Here all viral proteins required for cell-to-cell fusion are present in the cells, and the cells express the receptor CD46, but through altered localisation of the receptor, cell-to-cell fusion does not occur, allowing the persistent state to continue in these cells. The complexity of these processes indicates that our understanding of the replication of MV is still in its infancy.

References

Afzal MA, Osterhaus AD, Cosby SL, Jin L, Beeler J, Takeuchi K, Kawashima H (2003) Comparative evaluation of measles virus-specific RT-PCR methods through an international collaborative study. J Med Virol 70:171–176

Albertini AA, Wernimont AK, Muziol T, Ravelli RB, Clapier CR, Schoehn G, Weissenhorn W, Ruigrok RW (2006) Crystal structure of the rabies virus nucleoprotein-RNA complex. Science 313:360–363

Alkhatib G, Massie B, Briedis DJ (1988) Expression of bicistronic measles virus P/C mRNA by using hybrid adenoviruses: levels of C protein synthesized in vivo are unaffected by the presence or absence of the upstream P initiator codon. J Virol 62:4059–4069

Andzhaparidze OG, Chaplygina NM, Bogomolova NN, Lotte VD, Koptyaeva IB, Boriskin Y (1987) Non-infectious morphologically altered nucleocapsids of measles virus from persistently infected cells. Arch Virol 95:17–28

Auwaerter PG, Kaneshima H, McCune JM, Wiegand G, Griffin DE (1996) Measles virus infection of thymic epithelium in the SCID-hu mouse leads to thymocyte apoptosis. J Virol 70:3734–3740

Baczko K, Liebert UG, Billeter M, Cattaneo R, Budka H, ter Meulen V (1986) Expression of defective measles virus genes in brain tissues of patients with subacute sclerosing panencephalitis. J Virol 59:472–478

Banerjee AK (1987) Transcription and replication of rhabdoviruses 262. Microbiol Rev 51:66–87

Banerjee AK, Barik S, De BP (1991) Gene expression of nonsegmented negative strand RNA viruses. Pharmacol Ther 51:47–70

Bankamp B, Bellini WJ, Rota PA (1999) Comparison of L proteins of vaccine and wild-type measles viruses. J Gen Virol 80:1617–1625

Bankamp B, Wilson J, Bellini WJ, Rota PA (2005) Identification of naturally occurring amino acid variations that affect the ability of the measles virus C protein to regulate genome replication and transcription. Virology 336:120–129

Barr JN, Whelan SP, Wertz GW (2002) Transcriptional control of the RNA-dependent RNA polymerase of vesicular stomatitis virus. Biochim Biophys Acta 1577:337–353

Barrett T, Underwood B (1985) Comparison of messenger RNAs induced in cells infected with each member of the morbillivirus group. Virology 145:195–199

Bellini WJ, Englund G, Rozenblatt S, Arnheiter H, Richardson CD (1985) Measles virus P gene codes for two proteins. J Virol 53:908–919

Bellini WJ, Englund G, Richardson CD, Rozenblatt S, Lazzarini RA (1986) Matrix genes of measles virus and canine distemper virus: cloning, nucleotide sequences, and deduced amino acid sequences. J Virol 58:408–416

Bhella D, Ralph A, Yeo RP (2004) Conformational flexibility in recombinant measles virus nucleocapsids visualised by cryo-negative stain electron microscopy and real-space helical reconstruction. J Mol Biol 340:319–331

Blumberg BM, Crowley JC, Silverman JI, Menonna J, Cook SD, Dowling PC (1988) Measles virus L protein evidences elements of ancestral RNA polymerase. Virology 164:487–497

Bousse T, Matrosovich T, Portner A, Kato A, Nagai Y, Takimoto T (2002) The long noncoding region of the human parainfluenza virus type 1 F gene contributes to the read-through transcription at the M-F gene junction. J Virol 76:8244–8251

Brown DD, Rima BK, Allen IV, Baron MD, Banyard AC, Barrett T, Duprex WP (2005) Rational attenuation of a morbillivirus by modulating the activity of the RNA-dependent RNA polymerase. J Virol 79:14330–14338

Calain P, Roux L (1993) The rule of six, a basic feature for efficient replication of Sendai virus defective interfering RNA. J Virol 67:4822–4830

Campbell JJ, Cosby SL, Scott JK, Rima BK, Martin SJ, Appel M (1980) A comparison of measles and canine distemper virus polypeptides. J Gen Virol 48:149–159

Casali P, Sissons JG, Fujinami RS, Oldstone MB (1981) Purification of measles virus glycoproteins and their integration into artificial lipid membranes. J Gen Virol 54:161–171

Castaneda SJ, Wong TC (1990) Leader sequence distinguishes between translatable and encapsidated measles virus RNAs. J Virol 64:222–230

Cathomen T, Buchholz CJ, Spielhofer P, Cattaneo R (1995) Preferential initiation at the second AUG of the measles virus F mRNA: a role for the long untranslated region. Virology 214:628–632

Cattaneo R, Rebmann G, Schmid A, Baczko K, ter Meulen V, Billeter MA (1987) Altered transcription of a defective measles virus genome derived from a diseased human brain. EMBO J 6:681–688

Cattaneo R, Kaelin K, Baczko K, Billeter MA (1989a) Measles virus editing provides an additional cysteine-rich protein. Cell 56:759–764

Cattaneo R, Schmid A, Spielhofer P, Kaelin K, Baczko K, ter Meulen V, Pardowitz J, Flanagan S, Rima BK, Udem SA, Billeter MA (1989b) Mutated and hypermutated genes of persistent measles viruses which caused lethal human brain diseases. Virology 173:415–425

Cevik B, Smallwood S, Moyer SA (2003) The L-L oligomerization domain resides at the very N-terminus of the sendai virus L RNA polymerase protein. Virology 313:525–536

Chen M, Ogino T, Banerjee AK (2006) Mapping and functional role of the self-association domain of vesicular stomatitis virus phosphoprotein. J Virol 80:9511–9518

Chui LW, Vainionpaa R, Marusyk R, Salmi A, Norrby E (1986) Nuclear accumulation of measles virus nucleoprotein associated with a temperature-sensitive mutant. J Gen Virol 67:2153–2161

Curran J, Kolakofsky D (2008) Nonsegmented negative-strand RNA virus RNA synthesis in vivo. Virology 371:227–230

Curran JA, Richardson C, Kolakofsky D (1986) Ribosomal initiation at alternate AUGs on the Sendai virus P/C mRNA. J Virol 57:684–687

de Carvalho NC, Williamson RA, Parren PW, Lundkvist A, Burton DR, Björling E (2002) Neutralizing human Fab fragments against measles virus recovered by phage display. J Virol 76:251–258

de Swart RL, Ludlow M, de Witte L, Yanagi Y, van Amerongen G, McQuaid S, Yuksel S, Geijtenbeek TB, Duprex WP, Osterhaus AD (2007) Predominant infection of CD150[+] lymphocytes and dendritic cells during measles virus infection of macaques. PLoS Pathog 3:e178

Devaux P, von Messling V, Songsungthong W, Springfeld C, Cattaneo R (2007) Tyrosine 110 in the measles virus phosphoprotein is required to block STAT1 phosphorylation. Virology 360:72–83

Dowling PC, Blumberg BM, Menonna J, Adamus JE, Cook P, Crowley JC, Kolakofsky D, Cook SD (1986) Transcriptional map of the measles virus genome. J Gen Virol 67:1987–1992

Duprex WP, McQuaid S, Hangartner L, Billeter MA, Rima BK (1999) Observation of measles virus cell-to-cell spread in astrocytoma cells by using a green fluorescent protein-expressing recombinant virus. J Virol 73:9568–9575

Duprex WP, McQuaid S, Rima BK (2000a) Measles virus-induced disruption of the glial-fibrillary-acidic protein cytoskeleton in an astrocytoma cell line (U-251). J Virol 74:3874–3880

Duprex WP, McQuaid S, Roscic-Mrkic B, Cattaneo R, McCallister C, Rima BK (2000b) In vitro and in vivo infection of neural cells by a recombinant measles virus expressing enhanced green fluorescent protein. J Virol 74:7972–7979

Duprex WP, Collins FM, Rima BK (2002) Modulating the function of the measles virus RNA-dependent RNA polymerase by insertion of green fluorescent protein into the open reading frame. J Virol 76:7322–7328

Enders JF, Peebles TC (1954) Propagation in tissue cultures of cytopathic agents from patients with measles. Proc Soc Exp Biol Med 86:277–286

Ferron F, Longhi S, Henrissat B, Canard B (2002) Viral RNA-polymerases - a predicted 2'-O-ribose methyltransferase domain shared by all Mononegavirales. Trends Biochem Sci 27:222–224

Follett EA, Pringle CR, Pennington TH (1976) Events following the infections of enucleate cells with measles virus. J Gen Virol 32:163–175

Fontana JM, Bankamp B, Bellini WJ, Rota PA (2008) Regulation of interferon signaling by the C and V proteins from attenuated and wild-type strains of measles virus. Virology 374:71–81

Green TJ, Zhang X, Wertz GW, Luo M (2006) Structure of the vesicular stomatitis virus nucleo-protein-RNA complex. Science 313:357–360

Gubbay O, Curran J, Kolakofsky D (2001) Sendai virus genome synthesis and assembly are coupled: a possible mechanism to promote viral RNA polymerase processivity. J Gen Virol 82:2895–2903

Gupta KC, Kingsbury DW (1985) Polytranscripts of Sendai virus do not contain intervening polyadenylate sequences. Virology 141:102–109

Hall WW, ter Meulen V (1977) The effects of actinomycin D on RNA synthesis in measles virus-infected cells. J Gen Virol 34:391–396

Horikami SM, Moyer SA (1991) Synthesis of leader RNA and editing of the P mRNA during transcription by purified measles virus. J Virol 65:5342–5347

Huber M, Cattaneo R, Spielhofer P, Örvell C, Norrby E, Messerli M, Perriard JC, Billeter MA (1991) Measles virus phosphoprotein retains the nucleocapsid protein in the cytoplasm. Virology 185:299–308

Iizuka M, Smith MM (2003) Functional consequences of histone modifications. Curr Opin Genet Dev 13:154–160

Iseni F, Baudin F, Garcin D, Marq JB, Ruigrok RW, Kolakofsky D (2002) Chemical modification of nucleotide bases and mRNA editing depend on hexamer or nucleoprotein phase in Sendai virus nucleocapsids. RNA 8:1056–1067

Kingsbury DW (1974) The molecular biology of paramyxoviruses. Med Microbiol Immunol 160:73–83

Kozak M (1991a) A short leader sequence impairs the fidelity of initiation by eukaryotic ribosomes. Gene Expr 1:111–115

Kozak M (1991b) Effects of long 5' leader sequences on initiation by eukaryotic ribosomes in vitro. Gene Expr 1:117–125

Lamb RA, Parks GD (2007) *Paramyxoviridae*: the viruses and their replication. In: Knipe DM, Howley PM (eds) Fields Virology, 5th edn. Lippincott Williams Wilkins, Philadelphia, pp 1449–1496

Lamb RA, Mahy BW, Choppin PW (1976) The synthesis of Sendai virus polypeptides in infected cells. Virology 69:116–131

Liston P Briedis DJ (1994) Measles virus V protein binds zinc. Virology 198:399–404

Liu X, Bankamp B, Xu W, Bellini WJ, Rota PA (2006) The genomic termini of wild-type and vaccine strains of measles virus. Virus Res 122:78–84

Ludlow M, McQuaid S, Cosby SL, Cattaneo R, Rima BK, Duprex WP (2005) Measles virus superinfection immunity and receptor redistribution in persistently infected NT2 cells. J Gen Virol 86:2291–2303

Ludlow M, Duprex WP, Cosby SL, Allen IV, McQuaid S (2008) Advantages of using recombinant measles viruses expressing a fluorescent reporter gene with vibratome slice technology in experimental measles neuropathogenesis. Neuropathol Appl Neurobiol 34:424–434

Lund GA, Tyrrell DL, Bradley RD, Scraba DG (1984) The molecular length of measles virus RNA and the structural organization of measles nucleocapsids. J Gen Virol 65:1535–1542

Luo M, Green TJ, Zhang X, Tsao J, Qiu S (2007) Structural comparisons of the nucleoprotein from three negative strand RNA virus families. Virol J 4:72

Lyles DS, Rupprecht CE (2007) Rhabdoviridae. In: Knipe DM, Howley PM (eds) Fields Virology, 5th edn. Lippincott Williams Wilkins, Philadelphia, pp 1363–1408

McIlhatton MA, Curran MD, Rima BK (1997) Nucleotide sequence analysis of the large (L) genes of phocine distemper virus and canine distemper virus (corrected sequence) J Gen Virol 78:571–576

Nishie T, Nagata K, Takeuchi K (2007) The C protein of wild-type measles virus has the ability to shuttle between the nucleus and the cytoplasm. Microbes Infect 9:344–354

Ogura H, Baczko K, Rima BK, ter Meulen V (1987) Selective inhibition of translation of the mRNA coding for measles virus membrane protein at elevated temperatures. J Virol 61:472–479

Ogura H, Rima BK, Tas P, Baczko K, ter Meulen V (1988) Restricted synthesis of the fusion protein of measles virus at elevated temperatures. J Gen Virol 69:925–929

Pain VM (1996) Initiation of protein synthesis in eukaryotic cells. Eur J Biochem 236:747–771

Parks CL, Witko SE, Kotash C, Lin SL, Sidhu MS, Udem SA (2006) Role of V protein RNA binding in inhibition of measles virus minigenome replication. Virology 348:96–106

Paterson RG, Harris TJ, Lamb RA (1984) Analysis and gene assignment of mRNAs of a paramyxovirus, simian virus 5. Virology 138:310–323

Plumet S, Duprex WP, Gerlier D (2005) Dynamics of viral RNA synthesis during measles virus infection. J Virol 79:6900–6908

Plumet S, Herschke F, Bourhis JM, Valentin H, Longhi S, Gerlier D (2007) Cytosolic 5′-triphosphate ended viral leader transcript of measles virus as activator of the RIG I-mediated interferon response. PLoS ONE 2:e279

Poch O, Blumberg BM, Bougueleret L, Tordo N (1990) Sequence comparison of five polymerases (L proteins) of unsegmented negative-strand RNA viruses: theoretical assignment of functional domains. J Gen Virol 71:1153–1162

Portner A, Murti KG, Morgan EM, Kingsbury DW (1988) Antibodies against Sendai virus L protein: distribution of the protein in nucleocapsids revealed by immunoelectron microscopy. Virology 163:236–239

Qanungo KR, Shaji D, Mathur M, Banerjee AK (2004) Two RNA polymerase complexes from vesicular stomatitis virus-infected cells that carry out transcription and replication of genome RNA. Proc Natl Acad Sci U S A 101:5952–5957

Radecke F, Billeter MA (1995) Appendix: measles virus antigenome and protein consensus sequences. Curr Top Microbiol Immunol 191:181–192

Radecke F, Billeter (1996) The nonstructural C protein is not essential for multiplication of Edmonston B strain measles virus in cultured cells. Virology 217:418–421

Rager M, Vongpunsawad S, Duprex WP, Cattaneo R (2002) Polyploid measles virus with hexameric genome length. EMBO J 21:2364–2372

Rahaman A, Srinivasan N, Shamala N, Shaila MS (2004) Phosphoprotein of the rinderpest virus forms a tetramer through a coiled coil region important for biological function. A structural insight. J Biol Chem 279:23606–23614

Rennick LJ, Duprex WP, Rima BK (2007) Measles virus minigenomes encoding two autofluorescent proteins reveal cell-to-cell variation in reporter expression dependent on viral sequences between the transcription units. J Gen Virol 88:2710–2718

Richardson C, Hull D, Greer P, Hasel K, Berkovich A, Englund G, Bellini W, Rima B, Lazzarini R (1986) The nucleotide sequence of the mRNA encoding the fusion protein of measles virus

(Edmonston strain): a comparison of fusion proteins from several different paramyxoviruses. Virology 155:508–523

Richardson CD, Berkovich A, Rozenblatt S, Bellini WJ (1985) Use of antibodies directed against synthetic peptides for identifying cDNA clones, establishing reading frames, and deducing the gene order of measles virus. J Virol 54:186–193

Riddell MA, Rota JS, Rota PA (2005) Review of the temporal and geographical distribution of measles virus genotypes in the prevaccine and postvaccine eras. Virol J 2:87

Rima BK, Martin SJ (1979) Effect of undiluted passage on the polypeptides of measles virus. J Gen Virol 44:135–144

Rima BK, McFerran NV (1997) Dinucleotide and stop codon frequencies in single-stranded RNA viruses. J Gen Virol 78:2859–2870

Rima BK, Davidson WB, Martin SJ (1977) The role of defective interfering particles in persistent infection of Vero cells by measles virus. J Gen Virol 35:89–97

Rima BK, Baczko K, Clarke DK, Curran MD, Martin SJ, Billeter MA, ter Meulen V (1986) Characterization of clones for the sixth (L) gene and a transcriptional map for morbilliviruses. J Gen Virol 67:1971–1978

Rima BK, Earle JA, Yeo RP, Herlihy L, Baczko K, ter Meulen V, Carabana J, Caballero M, Celma ML, Fernandez-Munoz R (1995) Temporal and geographical distribution of measles virus genotypes. J Gen Virol 76:1173–1180

Rima BK, Earle JA, Baczko K, ter Meulen V, Liebert UG, Carstens C, Carabana J, Caballero M, Celma ML, Fernandez-Munoz R (1997) Sequence divergence of measles virus haemagglutinin during natural evolution and adaptation to cell culture. J Gen Virol 78:97–106

Rima BK, Collin AM, Earle JA (2005) Completion of the sequence of a cetacean morbillivirus and comparative analysis of the complete genome sequences of four morbilliviruses. Virus Genes 30:113–119

Rozenblatt S, Koch T, Pinhasi O, Bratosin S (1979) Infective substructures of measles virus from acutely and persistently infected cells. J Virol 32:329–333

Rudolph MG, Kraus I, Dickmanns A, Eickmann M, Garten W, Ficner R (2003) Crystal structure of the Borna disease virus nucleoprotein. Structure 11:1219–1226

Schneider-Schaulies S, Liebert UG, Baczko K, Cattaneo R, Billeter M, ter Meulen V (1989) Restriction of measles virus gene expression in acute and subacute encephalitis of Lewis rats. Virology 171:525–534

Schneider-Schaulies S, Kreth HW, Hofmann G, Billeter M, ter Meulen V (1991) Expression of measles virus RNA in peripheral blood mononuclear cells of patients with measles SSPE, and autoimmune diseases. Virology 182:703–711

Schneider-Schaulies S, Schneider-Schaulies J, Dunster LM, ter Meulen V (1995) Measles virus gene expression in neural cells. Curr Top Microbiol Immunol 191:101–116

Schneider H, Kaelin K, Billeter MA (1997) Recombinant measles viruses defective for RNA editing and V protein synthesis are viable in cultured cells. Virology 227:314–322

Schnorr JJ, Schneider-Schaulies S, Simon-Jodicke A, Pavlovic J, Horisberger MA, ter Meulen V (1993) MxA-dependent inhibition of measles virus glycoprotein synthesis in a stably transfected human monocytic cell line. J Virol 67:4760–4768

Schrag SJ, Rota PA, Bellini WJ (1999) Spontaneous mutation rate of measles virus: direct estimation based on mutations conferring monoclonal antibody resistance. J Virol 73:51–54

Sidhu MS, Chan J, Kaelin K, Spielhofer P, Radecke F, Schneider H, Masurekar M, Dowling PC, Billeter MA, Udem SA (1995) Rescue of synthetic measles virus minireplicons: measles genomic termini direct efficient expression and propagation of a reporter gene. Virology 208:800–807

Stallcup KC, Wechsler SL, Fields BN (1979) Purification of measles virus and characterization of subviral components. J Virol 30:166–176

Sugiyama T, Gursel M, Takeshita F, Coban C, Conover J, Kaisho T, Akira S, Klinman DM, Ishii KJ (2005) CpG RNA: identification of novel single-stranded RNA that stimulates human $CD14^+CD11c^+$ monocytes. J Immunol 174:2273–2279

Suryanarayana K, Baczko K, ter Meulen V, Wagner RR (1994) Transcription inhibition and other properties of matrix proteins expressed by M genes cloned from measles viruses and diseased human brain tissue. J Virol 68:1532–1543

Takeda M, Ohno S, Seki F, Nakatsu Y, Tahara M, Yanagi Y (2005) Long untranslated regions of the measles virus M and F genes control virus replication and cytopathogenicity. J Virol 79:14346–14354
Takeda M, Nakatsu Y, Ohno S, Seki F, Tahara M, Hashiguchi T, Yanagi Y (2006) Generation of measles virus with a segmented RNA genome. J Virol 80:4242–4248
Thorne HV, Dermott E (1977) Y-forms as possible intermediates in the replication of measles virus nucleocapsids. Nature 268:345–347
Udem SA, Cook KA (1984) Isolation and characterization of measles virus intracellular nucleocapsid RNA. J Virol 49:57–65
Valsamakis A, Schneider H, Auwaerter PG, Kaneshima H, Billeter MA, Griffin DE (1998) Recombinant measles viruses with mutations in the C, V, or F gene have altered growth phenotypes in vivo. J Virol 72:7754–7761
von Messling V, Cattaneo R (2002) Amino-terminal precursor sequence modulates canine distemper virus fusion protein function. J Virol 76:4172–4180
Walpita P (2004) An internal element of the measles virus antigenome promoter modulates replication efficiency. Virus Res 100:199–211
Washenberger CL, Han JQ, Kechris KJ, Jha BK, Silverman RH, Barton DJ (2007) Hepatitis C virus RNA: dinucleotide frequencies and cleavage by RNase L. Virus Res 130:85–95
Waterson A (1962) Two kinds of myxovirus. Nature 193:1163–1164
Whelan SP, Barr JN, Wertz GW (2004) Transcription and replication of nonsegmented negative-strand RNA viruses. Curr Top Microbiol Immunol 283:61–119
Wileman T (2007) Aggresomes and pericentriolar sites of virus assembly: cellular defense or viral design? Annu Rev Microbiol 61:149–167
Williams BR (1999) PKR: a sentinel kinase for cellular stress. Oncogene 18:6112–6120
Witko SE, Kotash C, Sidhu MS, Udem SA, Parks CL (2006) Inhibition of measles virus minireplicon-encoded reporter gene expression by V protein. Virology 348:107–119
Yoshikawa Y, Tsuruoka H, Matsumoto M, Haga T, Shioda T, Shibuta H, Sato TA, Yamanouchi K (1990) Molecular analysis of structural protein genes of the Yamagata-1 strain of defective subacute sclerosing panencephalitis virus. II. Nucleotide sequence of a cDNA corresponding to the P plus M dicistronic mRNA Virus Genes 4:151–161
Zuniga A, Wang Z, Liniger M, Hangartner L, Caballero M, Pavlovic J, Wild P, Viret JF, Glueck R, Billeter MA, Naim HY (2007) Attenuated measles virus as a vaccine vector. Vaccine 25:2974–2983

Chapter 6
Nucleocapsid Structure and Function

S. Longhi

Contents

The Replicative Complex of Measles Virus	104
Structural Disorder Within the N and P Proteins	107
Structural Organization of the Nucleoprotein	108
The Structured N_{CORE} Domain	109
The Intrinsically Unstructured N_{TAIL} Domain	111
Functional Role of Structural Disorder of N_{TAIL} for Transcription and Replication	115
N_{TAIL} and Molecular Partnership	120
Conclusions	122
References	123

Abstract Measles virus belongs to the *Paramyxoviridae* family within the Mononegavirales order. Its nonsegmented, single-stranded, negative-sense RNA genome is encapsidated by the nucleoprotein (N) to form a helical nucleocapsid. This ribonucleoproteic complex is the substrate for both transcription and replication. The RNA-dependent RNA polymerase binds to the nucleocapsid template via its co-factor, the phosphoprotein (P). This chapter describes the main structural information available on the nucleoprotein, showing that it consists of a structured core (N_{CORE}) and an intrinsically disordered C-terminal domain (N_{TAIL}). We propose a model where the dynamic breaking and reforming of the interaction between N_{TAIL} and P would allow the polymerase complex (L–P) to cartwheel on the nucleocapsid template. We also propose a model where the flexibility of the disordered N and P domains allows the formation of a tripartite complex (N°–P–L) during replication, followed by the delivery of N monomers to the newly synthesized genomic RNA chain. Finally, the functional implications of structural disorder are also discussed in light of the ability of disordered regions to establish interactions with multiple partners, thus leading to multiple biological effects.

S. Longhi
Architecture et Fonction des Macromolécules Biologiques, UMR 6098 CNRS et Universités Aix-Marseille I et II, 163 avenue de Luminy, Case 932, 13288 Marseille Cedex 09, France, e-mail: Sonia.Longhi@afmb.univ-mrs.fr

The Replicative Complex of Measles Virus

The genome of measles virus (MV) is encapsidated by the nucleoprotein (N) within a helical nucleocapsid. The viral RNA is tightly bound within the nucleocapsid and does not dissociate during RNA synthesis, as shown by its resistance to silencing by siRNA (Bitko and Barik 2001). Hence, this ribonucleoproteic (RNP) complex, rather than naked RNA, is the template for both transcription and replication. These latter activities are carried out by the RNA-dependent RNA polymerase (RdRp), which is composed of the large (L) protein and of the phosphoprotein (P). The P protein is an essential polymerase co-factor in that it tethers the L protein onto the nucleocapsid template. This ribonucleoproteic complex made of RNA, N, P, and L constitutes the replicative unit (Fig. 1A).

MV nucleocapsids, as visualized by negative stain transmission electron microscopy, have a typical herringbone-like appearance (Fig. 1B). The nucleocapsid of all *Paramyxoviridae* has a considerable conformational flexibility and can adopt different helical pitches (the axial rise per turn) and twists (the number of subunits per turn), resulting in conformations differing in their extent of compactness (Bhella et al. 2002, 2004; Egelman et al. 1989; Heggeness et al. 1980, 1981; Schoehn et al. 2004).

Once the viral RNPs are released in the cytoplasm of infected cells, transcription of viral genes occurs using endogenous NTPs as substrates. Following primary transcription, the polymerase switches to a processive mode and ignores the gene junctions to synthesize a full, complementary strand of genome length. This positive-stranded RNA (antigenome) does not serve as a template for transcription and its unique role is to provide an intermediate in replication. The intracellular concentration of the N protein is the main element controlling the relative rates of transcription and replication. When N is limiting, the polymerase is preferentially engaged in transcription, thus leading to an increase in the intracellular concentration of viral proteins, including N. When N levels are high enough to allow encapsidation of the nascent RNA chain, the polymerase switches to a replication mode (Plumet et al. 2005) (see also the chapter by B. Rima and W.P. Durprex, this volume and Albertini et al. 2005; Lamb and Kolakofsky 2001; Longhi and Canard 1999; Roux 2005 for reviews on transcription and replication).

The nucleoprotein is the most abundant structural protein. Its primary function is to encapsidate the viral genome. However, as we will see throughout this chapter, N is not a simple structural component, serving merely to package viral RNA. Rather, it plays several functions. Within *Mononegavirales*, each N monomer interacts with a precise number of nucleotides. The number of nucleotides varies among *Mononegavirales*, being specific to each family: six nucleotides for *Paramyxovirinae* (Egelman et al. 1989) and *Pneumovirinae* (Tran et al. 2007), nine for *Rhabdoviridae* (Albertini et al. 2006; Flamand et al. 1993; Green et al. 2006), and 12–15 for *Filoviridae* (Mavrakis et al. 2002). The fact that in *Paramyxovirinae* N wraps exactly six nucleotides imposes the so-called rule of six to these viruses, i.e., their genome must be of polyhexameric length (6n + 0) to efficiently replicate (see Kolakofsky et al. 2005; Roux 2005; Vulliemoz and Roux 2001 for reviews).

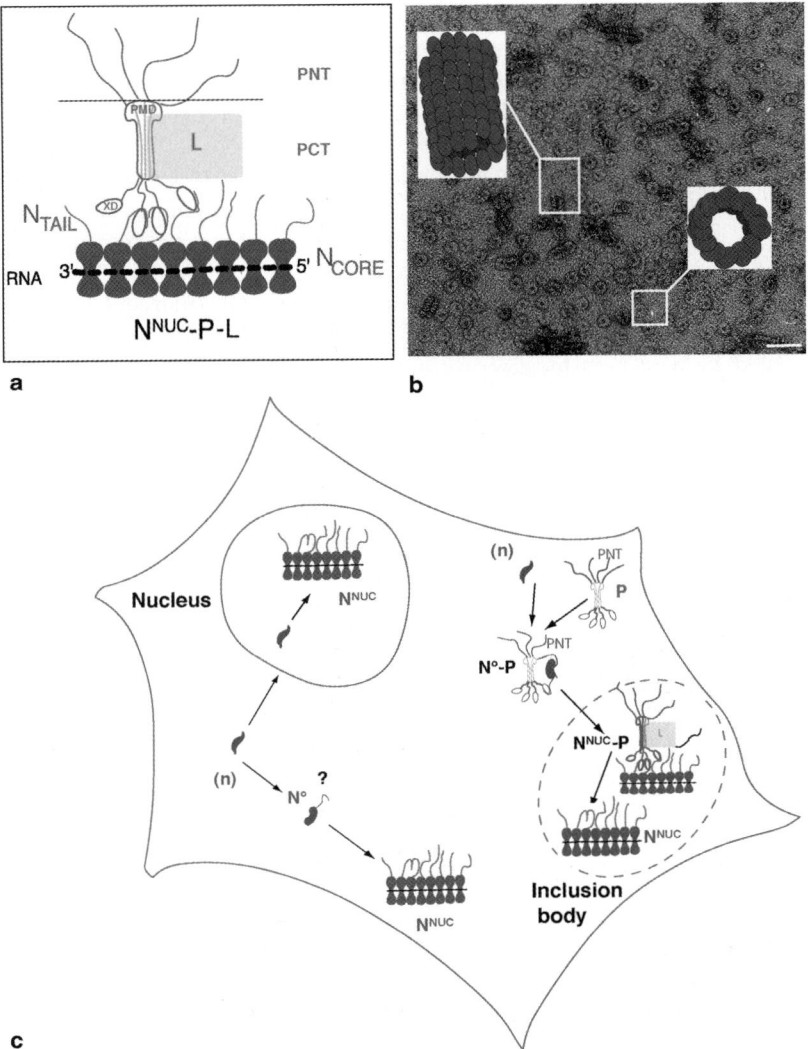

Fig. 1 A Schematic representation of the N^{NUC}–P–L complex of measles virus. The disordered N_{TAIL} (aa 401–525) and PNT (aa 1–230) regions are represented by lines. The encapsidated RNA is shown as a dotted line embedded in the middle of N by analogy with *Rhabdoviridae* N-RNA complexes (Albertini et al. 2006; Green et al. 2006). The multimerization domain of P (aa 304–375, PMD) is represented with a dumbbell shape according to Tarbouriech et al. (2000). The tetrameric P (Rahaman et al. 2004) is shown bound to N^{NUC} through three of its four C-terminal XD (aa 459–507) arms, as in the model of Curran and Kolakofsky (1999). The segment connecting PMD and XD is represented as disordered according to Longhi et al. (2003) and Karlin et al. (2003). The L protein is shown as a rectangle contacting P through PMD by analogy with SeV (Smallwood et al. 1994). **B** Negative-staining electron micrographs of bacterially expressed MV N. Nucleocapsid-like herringbone structures are shown on the left, while rings, corresponding to short nucleocapsids seen perpendicularly to their axis, are shown on the right. The bar represents 50 nm. Data shown in **B** were taken from Karlin et al. 2002a. **C** Maturation of MV N according to Gombart et al. (1993). Interior of an infected cell, with the nucleus (*top left*) and a cytoplasmic

The specific encapsidation signal is thought to lay within the 5′ *Leader* and *Trailer* extremities of the antigenome and genome RNA strands, respectively, as demonstrated for another *Mononegavirales* member (Blumberg et al. 1983). However, in the absence of viral RNA and of other viral proteins, *Mononegavirales* nucleoproteins are able to self-assemble onto cellular RNA to form nucleocapsid-like particles. Therefore, a regulatory mechanism is necessary to prevent the illegitimate self-assembly of N in the absence of ongoing genomic RNA synthesis. Indeed, in infected cells N is found in various forms: a soluble, monomeric form (referred to as N°) and an assembled form (referred to as N^{NUC}). Two forms of soluble N are likely to occur in the cytosol: a neosynthesized, transient form (herein referred to as n) and a more mature form, N° (Fig. 1C) (Gombart et al. 1993). The assembled form of N is localized in both the cytosol and the nucleus (Fig. 1C) (Gombart et al. 1993; Horikami and Moyer 1995). Sato and co-workers have recently identified the determinants of the intracellular trafficking within the nucleoprotein sequence of MV and canine distemper virus (CDV), a closely related *Morbillivirus*. They both possess a novel nuclear localization signal (NLS) at positions 70–77 and a nuclear export signal (NES). The NLS has a novel leucine/isoleucine-rich motif (TGILISIL), whereas the NES is composed of a leucine-rich motif (LLRSLTLF). While in CDV the NES occurs at positions 4–11, in MV it is located in the C-terminus (Sato et al. 2006). In both viruses, the nuclear export of N is CRM1-independent.

Once synthesized, the monomeric form of N requires the presence of a chaperone. This role is played by the P protein, whose association with N prevents illegitimate self-assembly of N and also retains the soluble form of N in the cytoplasm (Huber et al. 1991; Spehner et al. 1997). This soluble N°–P complex is used as a substrate for the encapsidation of nascent genomic RNA chain during replication. The assembled form of N also forms complexes with P, either isolated (N^{NUC}–P) or bound to L (N^{NUC}–P–L), which are essential to RNA synthesis by the viral polymerase (Buchholz et al. 1994; Ryan and Portner 1990) (Fig. 1C).

As the nucleoprotein, the P protein provides several functions in transcription and replication. Beyond serving as a chaperone for N, P binds to the nucleocapsid, thus tethering the polymerase onto the nucleocapsid template. P is a modular

---

Fig. 1 (Continued) inclusion body (*bottom right*). N is found under several conformations. The neosynthesized form (n) and a more mature form (N°) are both cytosolic, whereas the assembled form (N^{NUC}) is probably bound to the cytoskeleton (not shown). The polymerase complex is formed by L and by a tetramer of P. In the absence of P (*left*), n can self-assemble illegitimately on cellular RNA (*bottom*) and migrates to the nucleus, where it forms nucleocapsid-like particles. It is not known whether it undergoes a conformational change directly from n to N^{NUC} or whether N° is an intermediate conformation in the process. P forms a complex with N° (*right*) thereby preventing illegitimate self-assembly of N. The N°–P complex has been represented with a 1:4 stoichiometry by analogy to SeV (J. Curran, personal communication). Within the N°–P complex, the N_{CORE} region has been represented with a shape slightly differing from that of the N^{NUC}–P complex according to Gombart et al. (1993). The N°–P is used by the polymerase to encapsidate neosynthesized RNA during replication (which takes place in the cytoplasmic inclusion bodies). Reprinted with permission from Nova Publishers Inc.

protein, consisting of at least two domains: an N-terminal disordered domain (aa 1–230, PNT) (Karlin et al. 2002b) and a C-terminal domain (aa 231–507, PCT) (for a more detailed description of the modular organization of P see Bourhis et al. 2005a, 2006). Transcription requires only the PCT domain, whereas genome replication also requires PNT.

The viral polymerase, which is responsible for both transcription and replication, is poorly characterized. It is thought to carry out most (if not all) enzymatic activities required for transcription and replication, including nucleotide polymerization, mRNA capping, and polyadenylation. However, no *Paramyxoviridae* polymerase has been purified so far, implying that most of our present knowledge arises from bioinformatics studies. Notably, using bioinformatics approaches, a ribose-2'-O-methyltransferase domain involved in capping of viral mRNAs was identified within the C-terminal region of *Mononegavirales* polymerases (with the exception of *Bornaviridae* and *Nucleorhabdoviruses*) (Ferron et al. 2002). The methyltransferase activity of the C-terminal region of Sendai virus (SeV) polymerase (aa 1756–2228) has been recently demonstrated biochemically (Ogino et al. 2005). Interestingly, the polymerase of vesicular stomatitis virus (VSV, a *Rhabdoviridae* member) has been recently shown to possess both ribose-2'-O and guanine-N-7-methyltransferase activities (Li et al. 2006).

In all *Mononegavirales* members, the viral genomic RNA is always encapsidated by the N protein, and genomic replication does not occur in the absence of N°. Therefore, during RNA synthesis, the viral polymerase has to interact with the N:RNA complex and use the N°–P complex as the substrate of encapsidation. Hence, the components of the viral replication machinery – namely P, N, and L – engage in a complex macromolecular ballet. Although the understanding of the roles of N, P, and L within the replicative complex of MV has benefited from significant breakthroughs in recent years (see Bourhis et al. 2006 for a review), rather limited three-dimensional information on the replicative machinery is available. The scarcity of high-resolution structural data stems from several facts: (a) the problems in obtaining homogenous polymers of N suitable for X-ray analysis (Karlin et al. 2002a; Schoehn et al. 2001), (b) the low abundance of L in virions and its very large size, which renders its heterologous expression difficult, and (c) the structural flexibility of N and P (see below) (Bourhis et al. 2004, 2005b, 2006; Karlin et al. 2002b, 2003; Longhi et al. 2003).

Structural Disorder Within the N and P Proteins

In the course of the structural and functional characterization of MV replicative complex proteins, my group discovered that the N and P proteins contain long (up to 230 residues) disordered regions possessing sequence features that typify intrinsically disordered proteins (IDPs) (Bourhis et al. 2004, 2005a, 2005b, 2006; Karlin et al. 2003; Karlin et al. 2002b; Longhi et al. 2003). By using bioinformatics approaches (see Ferron et al. 2006), structural disorder was shown to be a conserved and widespread property within *Mononegavirales* N and P proteins, thus implying functional relevance (Karlin et al. 2003).

IDPs, also referred to as natively unfolded proteins, are functional proteins that fulfill essential biological functions while lacking highly populated constant secondary and tertiary structure under physiological conditions (Dunker et al. 2001; Dyson and Wright 2005; Fink 2005; Tompa 2003, 2005; Uversky 2002a; Uversky et al. 2005). Although there are IUPs that carry out their function while remaining disordered all the time (e.g., entropic chains) (Dunker et al. 2001), many of them undergo a disorder-to-order transition upon binding to their physiological partner(s), a process termed induced folding (Dyson and Wright 2002; Fuxreiter et al. 2004; Uversky 2002b).

The protein flexibility inherent in structural disorder is of functional relevance. In particular, an increased plasticity would (a) enable binding of numerous structurally distinct targets, (b) provide the ability to overcome steric restrictions, enabling larger interaction surfaces in protein–protein and protein–ligand complexes than those obtained with rigid partners, and (c) allow protein interactions to occur with both high specificity and low affinity (Dunker et al. 1998, 2001, 2002; Dunker and Obradovic 2001; Dyson and Wright 2005; Fink 2005; Gunasekaran et al. 2003; Uversky et al. 2002; Wright and Dyson 1999).

Regions lacking specific 3D structure have been so far associated with approximately 30 distinct functions, including nucleic acid and protein binding, display of phosphorylation sites and proteolysis sites, prevention of interactions by means of excluded volume effects, and molecular assembly. Most proteins containing disordered regions are involved in signaling and regulation events that generally imply multiple partner interactions (see Chen et al. 2006b; Dunker et al. 2002, 2005; Iakoucheva et al. 2002; Tompa 2003 for reviews). The percentage of the genome encoding protein disorder increases from bacteria to eukaryotes and, more generally, it increases with increasing organism complexity. The increased prevalence of disorder in higher organisms is related to an increased need for cell regulation and signaling. In addition, a recent study published by Dunker and Uversky's group shows that viruses and Eukaryota have ten times more conserved disorder (roughly 1%) than archaea and bacteria (0.1%) (Chen et al. 2006a). The abundance of disorder within viruses likely reflects the need for genetic compaction, where a single disordered protein can establish multiple interactions and hence exert multiple concomitant biological effects.

Structural Organization of the Nucleoprotein

Deletion analyses and electron microscopy studies have shown that *Paramyxoviridae* nucleoproteins are divided into two regions: a structured N-terminal moiety, N_{CORE} (aa 1–400 in MV), which contains all the regions necessary for self-assembly and RNA binding (Bankamp et al. 1996; Buchholz et al. 1993; Curran et al. 1993; Karlin et al. 2002a; Kingston et al. 2004b; Liston et al. 1997; Myers et al. 1997;

Myers et al. 1999), and a C-terminal domain, N_{TAIL} (aa 401–525 in MV), which is intrinsically disordered (Longhi et al. 2003) (Fig. 2A). N_{TAIL} protrudes from the globular body of N_{CORE} and is exposed at the surface of the viral nucleocapsid (Heggeness et al. 1980, 1981; Karlin et al. 2002a). N_{TAIL} contains the regions responsible for binding to P in both N°–P and N^{NUC}–P complexes (Bankamp et al. 1996; Kingston et al. 2004b; Liston et al. 1997; Longhi et al. 2003).

The Structured N_{CORE} Domain

N_{CORE} contains all the regions necessary for self-assembly and RNA binding, since nucleoproteins composed only of the core region can encapsidate neosynthesized RNA into nucleocapsid-like particles. Within N_{CORE}, deletion studies have failed to identify independent, modular domains, but have identified regions involved in the N–N interaction. The region spanning aa 258–357, called the central conserved region (CCR), is well conserved in sequence, and mainly hydrophobic (Fig. 2A). Several studies have shown that it is one of the regions involved in self-association and in RNA binding (Karlin et al. 2002a; Liston et al. 1997). The location of the RNA-binding site(s) within N_{CORE} is unknown. When denatured, such as in Northwestern blots, N does not bind RNA, indicating that the RNA-binding site is probably formed by maturation of N during encapsidation (Lamb and Kolakofsky 2001). In agreement, all mutants in which self-association is impaired do not package RNA (Karlin et al. 2002a; Myers et al. 1999). In the closely related SeV (a *Respirovirus* member), more subtle mutations in the 360–375 region were found that disrupted neither RNA binding nor the morphology of N, but rendered N inactive in replication (Myers et al. 1999). The author suggested that particular residues in this region might be involved in binding the leader sequence of SeV RNA, as opposed to nonspecific RNA. However, it is more likely that mutations within this region could affect the ability of N to interact with PNT, thereby preventing either formation or proper conformation of the N°–P complex.

Beyond the CCR, another region of N–N interaction was found within the 189–239 residues of MV N (see Fig. 2A), as deletion of this region led to a nucleoprotein variant form that can still bind P but has lost its ability to self-assemble (Bankamp et al. 1996). In further support of a role of this additional region in N assembly, two point substitutions thereof (namely, S228Q and L229D) impair self-association of N and RNA binding without affecting either the overall secondary structure content or the gross domain organization and the ability to bind P (Karlin et al. 2002a).

Indeed, a major hurdle to X-ray crystallography techniques is the strong self-assembly of N to form large nucleocapsids with a broad size distribution when expressed in heterologous systems such as mammalian cells (Spehner et al. 1991), bacteria (Warnes et al. 1995) and insect cells (Bhella et al. 2002). Because of this property, *Paramyxoviridae* nucleoproteins have resisted high-resolution structure determination so far. Because of variable helical parameters, the recombinant or

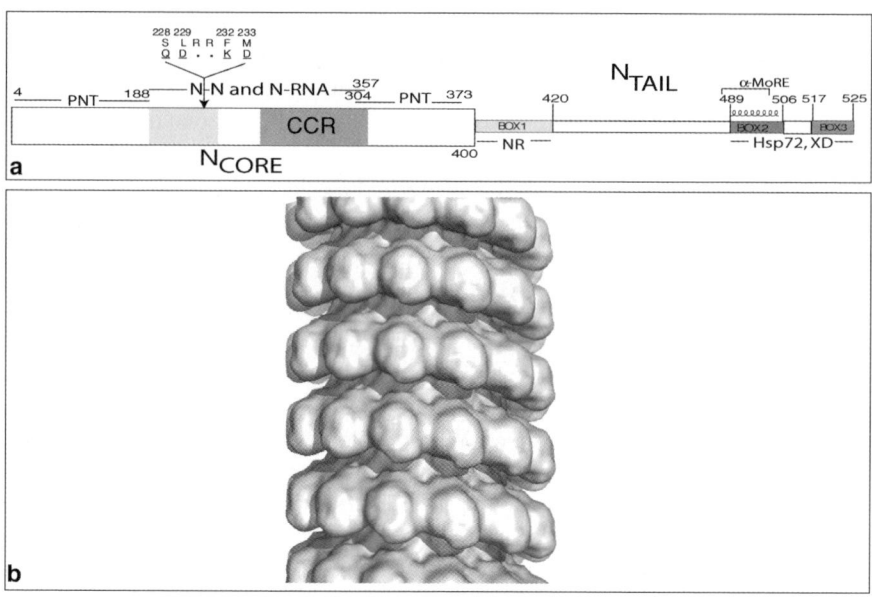

Fig. 2 A Organization of MV N. The location of N-N, N-P, and RNA binding sites is indicated. The central conserved region (CCR, aa 258–357) (*dark gray*) and the additional region involved in oligomerization (aa 189–239; *light gray*) are also shown. The positions targeted for mutagenesis (see text) are shown by an *arrow*. Wild-type residues are shown in the *top position*, while mutated residues are shown *below*. The location of the N_{CORE}-PNT sites (aa 4–188 and 304–373), as reported by Bankamp et al. (1996), is also shown. **B** Reconstruction of MV nucleocapsid as obtained from cryoelectron microscopy. Data were kindly provided by David Bhella (MRC, Glasgow, UK). The picture was drawn using Pymol (DeLano, W.L. The PyMOL Molecular Graphics System (2002) DeLano Scientific, San Carlos, CA, USA. http://www.pymol.org). (Reprinted with permission from Nova Publishers Inc.)

viral nucleocapsids are also difficult to analyze using electron microscopy coupled to image analysis. Despite these technical drawbacks, elegant electron microscopy studies by two independent groups led to real-space helical reconstruction of MV nucleocapsids (Bhella et al. 2004; Schoehn et al. 2004) (Fig. 2B). These studies pointed out a considerable conformational flexibility, with the most extended conformation having a helical pitch of 66 Å, while twist varies from 13.04 to 13.44 (Bhella et al. 2004). Notably, these studies also highlighted cross-talk between N_{CORE} and N_{TAIL}, based on the observation that removal of the disordered N_{TAIL} domain leads to increased nucleocapsid rigidity, with significant changes in both pitch and twist (see Bhella et al. 2004; Longhi et al. 2003; Schoehn et al. 2004).

Conversely, high-resolution structural data are available for two *Rhabdoviridae* members, namely the rabies virus and the VSV (Albertini et al. 2006; Green et al. 2006; Luo et al. 2007). The nucleoprotein of these viruses consists of two lobes and possesses an extended terminal arm that makes contacts with a neighboring N monomer. The RNA is tightly packed in a cavity between the two N lobes. N establishes contacts with the sugar and phosphate moiety of nucleotides via basic residues, in agreement with previous studies showing that the phosphate moieties of encapsidated RNA are not accessible to the solvent (Iseni et al. 2000). In both nucleoproteins, the RNA is not accessible to the solvent. Thus, it has to partially dissociate from N to become accessible to the polymerase.

Functional and structural similarities between the nucleoproteins of *Rhabdoviridae* and *Paramyxoviridae* are well established. In particular, they share the same organization in two well-defined regions, N_{CORE} and N_{TAIL}, and in both families the CCR is involved in RNA binding and self-assembly of N (Kouznetzoff et al. 1999). Moreover, incubation of the rabies virus nucleocapsid with trypsin results in the removal of the C-terminal region (aa 377–450) (Kouznetzoff et al. 1998). This N_{TAIL}-free nucleocapsid is no longer able to bind to P, thus suggesting that in *Rhabdoviridae*, N_{TAIL} plays a role in the recruitment of P, like in *Paramyxoviridae* (Schoehn et al. 2001). However, contrary to MV, the rabies virus N_{TAIL} domain is structured (Albertini et al. 2006). Presently, it is not known whether *Rhabdoviridae* and *Paramyxoviridae* nucleoproteins share the same bilobal morphology. In MV N, bioinformatics analyses predict that N_{CORE} is organized into two subdomains (aa 1–130 and aa 145–400) (see Bourhis et al. 2007; Ferron et al. 2006) separated by a hypervariable, antigenic loop (aa 131–149) (Giraudon et al. 1988), that would probably fold cooperatively into a bilobal morphology.

Although within the MV N^{NUC}–P complex, the N region responsible for binding to P is located within N_{TAIL}, the N°–P complex involves an additional interaction between N_{CORE} and the disordered N-terminal domain of P (PNT) (Fig. 1C). Within the N°–P complex, P-to-N binding is mediated by the dual PNT–N_{CORE} and PCT–N_{TAIL} interaction (Chen et al. 2003) (Fig. 1C). Studies on SeV suggested that an N°–P complex is absolutely necessary for the polymerase to initiate encapsidation (Baker and Moyer 1988). Therefore, formation of the N°–P complex would have at least two separate functions: (a) preventing illegitimate self-assembly of N and (b) allowing the polymerase to deliver N to the nascent RNA to initiate replication.

The regions within N_{CORE} responsible for binding PNT within the N°–P complex have been mapped to residues 4–188 and 304–373, with the latter region being not strictly required for binding but rather favoring it (Bankamp et al. 1996) (Fig. 2A). However, precise mapping of such regions is hard because N_{CORE} does not have a modular structure, and consequently it is difficult to distinguish between gross structural defects and specific effects of deletions.

The Intrinsically Unstructured N_{TAIL} Domain

In *Morbilliviruses*, N_{TAIL} is responsible for binding to P in both N°–P and N^{NUC}–P complexes (Bankamp et al. 1996; Kingston et al. 2004; Liston et al. 1997; Longhi et al. 2003). Within the N^{NUC}–P complex, N_{TAIL} is also responsible for the interaction with the polymerase (L–P) complex (Bankamp et al. 1996; Kingston et al. 2004; Liston et al. 1997; Longhi et al. 2003). Several features distinguish N_{TAIL} from N_{CORE}. Indeed, N_{TAIL} possesses features that are hallmarks of intrinsic disorder:

(a) it is hypersensitive to proteolysis (Karlin et al. 2002a), (b) it is not visualizable in cryoelectron microscopy reconstructions of nucleocapsids (Bhella et al. 2004), and (c) it has an amino acid sequence that is hypervariable among *Morbillivirus* members.

In agreement with these features suggesting disorder, the sequence properties of N_{TAIL} conform to those of IDPs (Dunker et al. 2001). Indeed, while the amino acid composition of N_{CORE} does not deviate from the average composition of proteins found in the Protein Data Bank (PDB), the N_{TAIL} region is depleted in order-promoting residues (W, C, F, Y, I, L) and enriched in disorder-promoting residues (R, Q, S, and E) (Longhi et al. 2003). Moreover, N_{TAIL} is predicted to be mainly (if not fully) disordered (data not shown; Bourhis et al. 2004; Longhi et al. 2003) by the secondary structure and disorder predictors implemented within the MeDor metaserver (Lieutaud et al., in press).

The disordered state of N_{TAIL} was experimentally confirmed by several spectroscopic and hydrodynamic approaches (see Receveur-Bréchot et al. 2006 for a review on methods to assess structural disorder and induced folding). Altogether, these studies pointed out that N_{TAIL} is a premolten globule (Bourhis et al. 2004; Longhi et al. 2003), i.e., it has a conformational state intermediate between a random coil and a molten globule (Dunker et al. 2001; Uversky 2002a). In solution, premolten globules possess a certain degree of residual compactness due to the presence of residual and fluctuating secondary and tertiary structures. As for the functional implications, it has been proposed that the residual intramolecular interactions that typify the premolten globule state may enable a more efficient start of the folding process induced by a partner (Fuxreiter et al. 2004; Lacy et al. 2004; Tompa 2002).

That N_{TAIL} does indeed undergo induced folding was documented by CD studies, where N_{TAIL} was shown to undergo an α-helical transition in the presence of PCT (Longhi et al. 2003). Using computational approaches, an α-helical *M*olecular *R*ecognition *E*lement (α-MoRE, aa 488–499 of N) has been identified within N_{TAIL} (see Fig. 2A). MoREs are regions within IDPs that have a certain propensity to bind to a partner and thereby undergo induced folding (Garner et al. 1999; Mohan et al. 2006; Oldfield et al. 2005; Vacic et al. 2007). The role of the α-MoRE in binding to P and in the α-helical induced folding has further been confirmed by spectroscopic and biochemical experiments carried out on a truncated N_{TAIL} form devoid of the 489–525 region (Bourhis et al. 2004).

The P region responsible for the interaction with N_{TAIL} and its induced folding has been mapped to the C-terminal module (XD, aa 459–507; see Fig. 1A) of P (Johansson et al. 2003). The crystal structure of XD has been solved and consists of a triple α-helical bundle. A model of the interaction between XD and the α-MoRE of N_{TAIL} was then built in which N_{TAIL} is embedded in a large XD hydrophobic cleft delimited by helices α2 and α3 of XD. According to this model, burying of hydrophobic residues of the α-MoRE would provide the driving force to induce its folding, thus leading to a pseudo-four-helix arrangement occurring frequently in nature (Johansson et al. 2003). This model was thereafter validated by Kingston and co-workers who solved the crystal structure of a chimeric form mimicking this complex (Fig. 3) (Kingston et al. 2004).

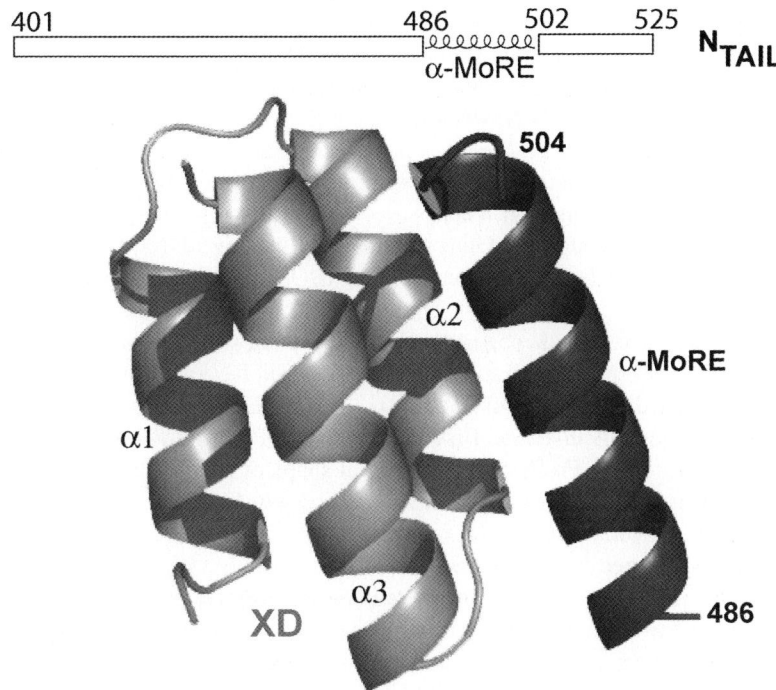

Fig. 3 Ribbon representation of the crystal structure of the complex between XD and the 486–504 region of N_{TAIL} (pdb code: 1T60) (Kingston et al. 2004). The picture was drawn using Pymol (DeLano, W.L. The PyMOL Molecular Graphics System (2002) DeLano Scientific, San Carlos, CA, USA, http://www.pymol.org). The schematic representation of N_{TAIL} with the α-MoRE highlighted is shown on the top

Small-angle X-ray scattering studies provided a low resolution model of the N_{TAIL}–XD complex, which showed that most of N_{TAIL} (aa 401–488) remains disordered within the complex and supported a role for the C-terminus in binding to XD (Bourhis et al. 2005). The involvement of the C-terminus of N_{TAIL} in binding to XD was indeed confirmed by spectroscopic and surface plasmon resonance (BIAcore) studies, where removal of either Box3 alone or Box2 plus Box3 (see Fig. 2A) results in a strong increase (three orders of magnitude, 10µM K_D) in the equilibrium dissociation constant (Bourhis et al. 2005). When synthetic peptides mimicking Box1, Box2, and Box3 were used, Box2 peptide was found to display an affinity toward XD (20nM K_D) similar to that of N_{TAIL} (80nM K_D), consistent with the role of Box2 as the primary binding site (S. Longhi and M.J. Oglesbee, unpublished data). Interestingly however, Box3 peptide exhibits an insignificant affinity for XD (approximately 1mM K_D) (S. Longhi and M.J. Oglesbee, unpublished data). The discrepancy between the data obtained with N_{TAIL} truncated proteins and with peptides can be accounted for by assuming that Box3 would act only in the context of N_{TAIL} and not in isolation. Thus, according to this model, Box3 and Box2 would be functionally coupled in the binding of N_{TAIL} to XD. One can speculate that burying the hydrophobic side of the α-MoRE in the hydrophobic cleft formed by helices α2 and α3 of XD would provide the primary driving force in the N_{TAIL}-XD interaction and that Box3 would act by stabilizing the bound conformation.

Heteronuclear NMR (HN-NMR) studies using ^{15}N-labeled N_{TAIL} or a truncated form devoid of Box3 and unlabeled XD, revealed that while Box2 undergoes α-helical folding upon binding to XD, Box3 does not acquire any regular secondary structure element (Bourhis et al. 2005).

The molecular mechanism of the XD-induced folding of N_{TAIL} has been also investigated by using site-directed spin-labeling (SDSL) electron paramagnetic resonance (EPR) spectroscopy. The basic strategy of SDSL involves the introduction of a paramagnetic nitroxide side chain at a selected protein site. This is usually accomplished by cysteine-substitution mutagenesis, followed by covalent modification of the unique sulphydryl group with a selective nitroxide reagent, such as the methanethiosulphonate derivative (see Biswas et al. 2001; Feix and Klug 1998; Hubbell et al. 1998 for reviews). From the EPR spectral shape of a spin-labeled protein, one can extract information in terms of radical mobility, which reflects the local mobility of residues in the proximity of the radical. Variations in the radical mobility can therefore be monitored in the presence of partners, ligands, or organic solvents.

Fourteen single-site N_{TAIL} cysteine mutants were designed, purified, and labeled, thus enabling grafting of a nitroxide paramagnetic probe on 12 sites scattered in the 488–525 region and on two sites located outside the reported region of interaction with XD (Morin et al. 2006; V. Belle et al., in press). EPR spectra were then recorded in the presence of either the secondary structure stabilizer 2,2,2-trifluoroethanol (TFE) or XD.

Different regions of N_{TAIL} were shown to contribute to a different extent to the binding to XD: while the mobility of the spin labels grafted at positions 407 and 460 was unaffected upon addition of XD, that of the spin labels grafted within the 488–502 and the 505–522 regions was severely and moderately reduced, respectively. Furthermore, EPR experiments in the presence of 30% sucrose (i.e., under conditions in which the intrinsic motion of the protein becomes negligible with respect to the intrinsic motion of the spin label), allowed precise mapping of the N_{TAIL} region undergoing α-helical folding to residues 488–502. The drop in the mobility of the 505–522 region upon binding to XD was shown to be comparable to that observed in the presence of TFE, thus suggesting that the restrained mobility that this region experiences upon binding to XD is not due to a direct interaction with XD.

The mobility of the 488–502 region was found to be restrained even in the absence of the partner, a behavior that could be accounted for by the existence of a transiently populated folded state. This may reflect the predominance of an α-helical conformation among the highly fluctuating conformations sampled by unbound N_{TAIL}, in agreement with previous biochemical and spectroscopic data that mapped the region involved in the α-helical-induced folding to residues 489–506 (Bourhis et al. 2004). That the conformational space of MoREs (Oldfield et al. 2005) in the unbound state is restricted by their inherent conformational propensities, thereby reducing the entropic cost of binding, has already been proposed (Fuxreiter et al. 2004; Lacy et al. 2004; Sivakolundu et al. 2005; Tompa 2002).

Finally, equilibrium displacement experiments showed that the XD-induced folding of N_{TAIL} is a reversible phenomenon (Morin et al. 2006). These results represent the first experimental evidence indicating that N_{TAIL} adopts its original premolten globule conformation after dissociation from its partner. This latter point is particularly relevant taking into consideration that the contact between XD and N_{TAIL} within the replicative complex has to be dynamically made and broken to allow the polymerase to progress along the nucleocapsid template during both transcription and replication. Hence, the complex cannot be excessively stable for this transition to occur efficiently at a high rate.

In conclusion, using a panel of various physicochemical approaches, the interaction between N_{TAIL} and XD was shown to imply the stabilization of the helical conformation of the α-MoRE, which is otherwise only transiently populated in the unbound form. The occurrence of a transiently populated α-helix, even in the absence of the partner, suggests that the molecular mechanism governing the folding of N_{TAIL} induced by XD may rely on conformer selection (i.e., selection by the partner of a preexisting conformation) (Tsai et al. 2001a, 2001b) rather than on a fly-casting mechanism (Shoemaker et al. 2000), contrary to what has been reported for the pKID–KIX couple (Sugase et al. 2007).

Stabilization of the helical conformation of the α-MoRE is also accompanied by a reduction in the mobility of the downstream region. The lower flexibility of the region downstream Box2 is not caused by a gain of α-helicity, nor can it be ascribed to a restrained motion brought by a direct interaction with XD. Rather, it likely arises from a gain of rigidity brought by α-helical folding of the neighboring Box2 region.

In agreement with these data and with the BIAcore data obtained using a synthetic Box3 peptide, preliminary titration studies using heteronuclear NMR suggest that XD does not establish direct interactions with Box3 (H. Darbon and S. Longhi, unpublished data). A tentative model can be proposed, where binding to XD might take place through a sequential mechanism that could involve binding and α-helical folding of Box2, followed by a conformational change of Box3, whose overall mobility is consequently reduced, probably through tertiary contacts with the neighboring Box2 region.

Functional Role of Structural Disorder of N_{TAIL} for Transcription and Replication

The K_D value between N_{TAIL} and XD is in the 100nM range (Bourhis et al. 2005). This affinity is considerably higher than that derived from isothermal titration calorimetry studies which pointed out a KD of 13μM (Kingston et al. 2004). A weak binding affinity, implying fast association and dissociation rates, would ideally fulfill the requirements of a polymerase complex that has to cartwheel on the nucleocapsid template during both transcription and replication. However, a K_D in the micromolar range would not seem to be physiologically relevant considering the low intracellular concentrations of P in the early phases of infection and the relatively long half-life of active P–L transcriptase complex tethered on the NC template, which has been determined to be well over 6 h (Plumet et al. 2005). Moreover, such a weak affinity is not consistent with the ability to readily purify nucleocapsid–P complexes using rather stringent techniques such as CsCl isopycnic density centrifugation (Oglesbee et al. 1989; Robbins and Bussell 1979; Robbins et al. 1980; Stallcup et al. 1979). A more stable XD–N_{TAIL} complex would be predicted to hinder the processive movement of P along the nucleocapsid template. In agreement with this model, the elongation rate of MV polymerase was found to be rather slow (three nucleotides/s) (Plumet et al. 2005). In addition, the C-terminus of N_{TAIL} has been shown to have an inhibitory role upon transcription and replication, as indicated by minireplicon experiments, where deletion of the C-terminus of N enhanced basal reporter gene expression (Zhang et al. 2002). Deletion of the C-terminus of N also reduces the affinity of XD for N_{TAIL}, providing further support for modulation of XD/N_{TAIL} binding affinity as a basis for polymerase processivity. Thus, Box3 would dynamically control the strength of the N_{TAIL}-XD interaction, by stabilizing the Box2–XD interaction. Removal of Box3 or interaction of Box3 with other partners (see the next paragraph) would reduce the affinity of N_{TAIL} for XD, thus stimulating transcription. Modulation of XD/N_{TAIL} binding affinity could be dictated by interactions between N_{TAIL} and cellular and/or viral co-factors. Indeed, the requirement for cellular or viral co-factors in both transcription and replication has been already documented in MV (Vincent et al. 2002) and other *Mononegavirales* members (Fearns and Collins 1999; Hartlieb et al. 2003). Furthermore, in both CDV and MV, viral transcription and replication are enhanced by the major heat shock protein (hsp72), and this stimulation relies on interaction with N_{TAIL} (Zhang et al. 2002, 2005). These co-factors may serve as processivity or transcription elongation factors and could act by modulating the strength of the interaction between the polymerase complex and the nucleocapsid template (see the next paragraph).

N_{TAIL} also influences the physical properties of the nucleocapsid helix that is formed by N_{CORE} (Longhi et al. 2003; Schoehn et al. 2004). Electron microscopic analysis of nucleocapsids formed by either N or N_{CORE} indicates that the presence of N_{TAIL} was associated with a greater degree of fragility, evidenced by the tendency of helices to break into individual ring structures (Fig. 4A). This fragility is associated with evidence of increased nucleocapsid flexibility, with helices formed by N_{CORE} alone forming rods (Fig. 4A) (see also Schoehn et al. 2004). It is therefore conceivable that the induced folding of N_{TAIL} resulting from the interaction with P (and/or other physiological partners) could also affect the nucleocapsid conformation in such a way as to affect the structure of the replication promoter (Fig. 4B). Indeed, the replication promoter, located at the 3′ end of the viral genome, is composed of two discontinuous elements building up a functional unit because of their juxtaposition on two successive helical turns (Tapparel et al. 1998) (Fig. 4B). The switch between transcription and replication could be dictated by variations in the helical conformation of the nucleocapsid, which would result in a modification in the number of N monomers (and thus of nucleotides) per turn, thereby disrupting the replication promoter in favor of the transcription promoter (or vice-versa).

6 Nucleocapsid Structure and Function

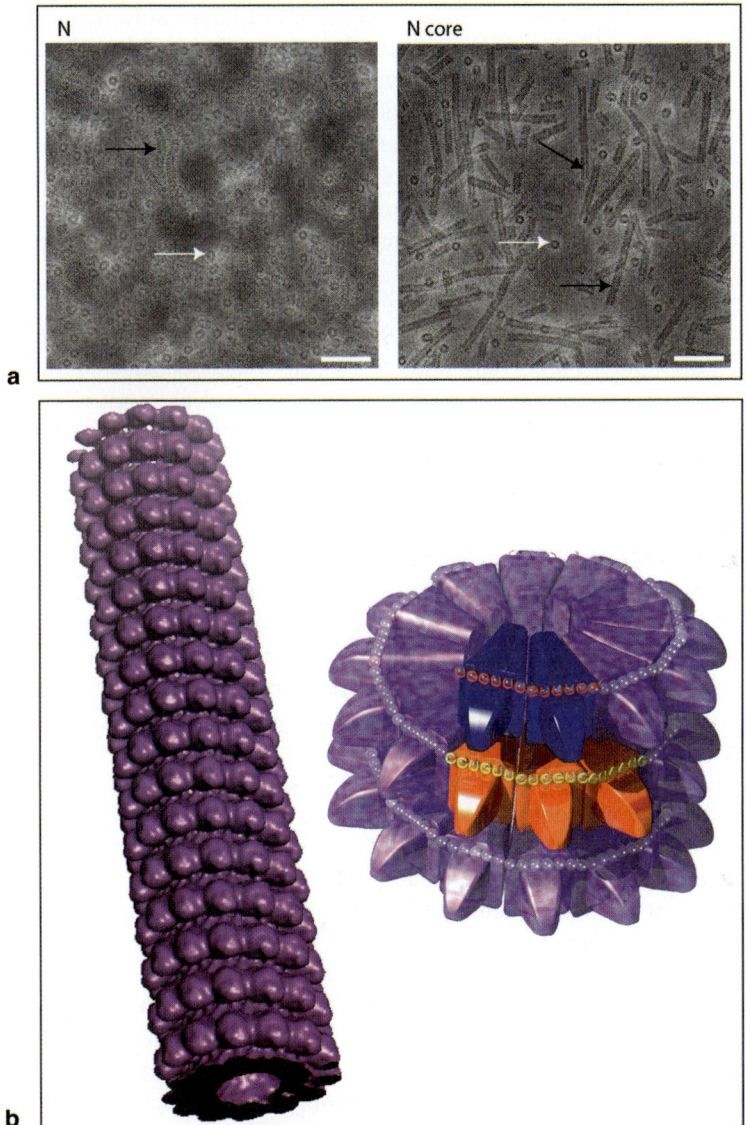

Fig. 4 A Negative stain electron micrographs of N and N_{CORE}. The bar corresponds to 100 nm. Rings and herringbone structures are indicated by *white and black arrows*, respectively. **B** Cryoelectron microscopy reconstructions of MV nucleocapsid (*left*) and schematic representation of the nucleocapsid (*right*), highlighting the structure of the replication promoter composed of two discontinuous units juxtaposed on successive helical turns (see regions wrapped by the *red* and *blue* N monomers). (Courtesy of D. Bhella, MRC, Glasgow, UK). Data in **A** were taken from Longhi et al. (2003). Reprinted with permission from Nova Publishers Inc.

Morphological analyses, showing the occurrence of a large conformational flexibility within *Paramyxoviridae* nucleocapsids (Bhella et al. 2002, 2004; Oglesbee et al. 1989, 1990), tend to corroborate this hypothesis.

Finally, preliminary data indicate that incubation of MV NCs in the presence of XD triggers unwinding of the NC, thus possibly enhancing the accessibility of genomic RNA to the polymerase complex (D. Bhella and S. Longhi, unpublished data). Hence, it is tempting to propose that the XD-induced α-helical folding of N_{TAIL} could trigger the opening of the two lobes of N_{CORE}, thus rendering the genomic RNA accessible to the solvent.

Unstructured regions are considerably more extended in solution than globular regions. For instance, MV PNT has a Stokes radius of 4 nm (Karlin et al. 2002). However, the Stokes radius only reflects a mean dimension. Indeed, the maximal extension of PNT, as measured by SAXS (S. Longhi et al., unpublished data) is considerably larger (>40 nm). In comparison, one turn of the nucleocapsid is 18 nm in diameter and 6 nm high (Bhella et al. 2002). Thus PNT could easily stretch over several turns of the nucleocapsid, and since P is multimeric, N°–P might have a considerable extension (Fig. 5). In the same vein, it is striking that SeV and MV PCT, which interacts with the intrinsically disordered N_{TAIL} domain, comprises a flexible linker (Bernado et al. 2005; Blanchard et al. 2004; Longhi et al. 2003; Marion et al. 2001). This certainly suggests the need for a great structural flexibility. This flexibility could be necessary for the tetrameric P to bind several turns of the helical nucleocapsid. Indeed, the promoter signals for the polymerase are located on the first and the second turn of the SeV nucleocapsid (Tapparel et al. 1998).

Likewise, the maximal extension of N_{TAIL} in solution is of 13 nm (Longhi et al. 2003). The very long reach of disordered regions could enable them to act as linkers and to tether partners on large macromolecular assemblies. Accordingly, one role of the tentacular N_{TAIL} projections in actively replicating nucleocapsids could be to put into contact several proteins within the replicative complex, such as the N°–P and the P–L complexes (see Fig. 5).

In *Respiroviruses* and *Morbilliviruses*, PNT, which is also disordered (Karlin et al. 2003), contains binding sites for N° (Curran et al. 1994; Harty and Palese 1995; Sweetman et al. 2001) and for L (Curran et al. 1995; Curran and Kolakofsky 1991; Sweetman et al. 2001). This pattern of interactions among N°, P, and L, mediated by unstructured regions of either P or N, suggests that N°, P, and L might interact simultaneously at some point during replication. Notably, the existence of a N–P–L tripartite complex has been proved by co-immunoprecipitation studies in the case of VSV, where this tripartite complex constitutes the replicase complex, as opposed to the L–P binary transcriptase complex (Gupta et al. 2003).

A model can be proposed where during replication, the extended conformation of PNT and N_{TAIL} would be key to allowing contact between the assembly substrate (N°–P) and the polymerase complex (L–P), thus leading to a tripartite N°–P–L complex (Fig. 5). This model emphasizes the plasticity of intrinsically disordered regions, which might give a considerable reach to the elements of the replicative machinery.

6 Nucleocapsid Structure and Function

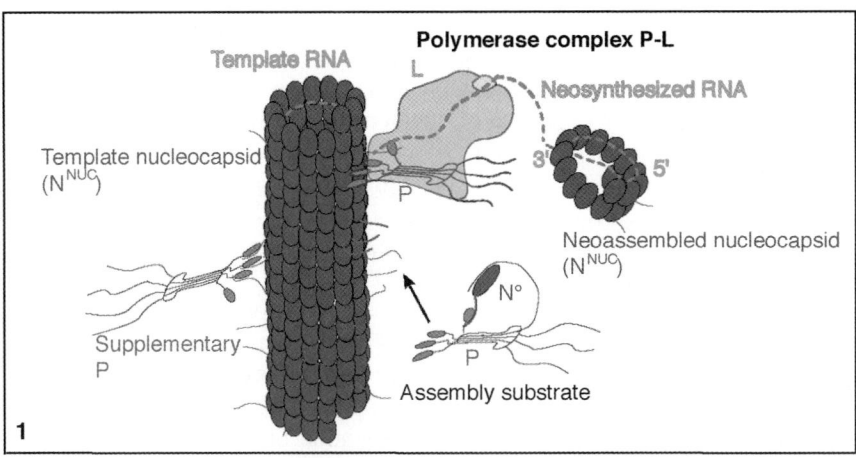

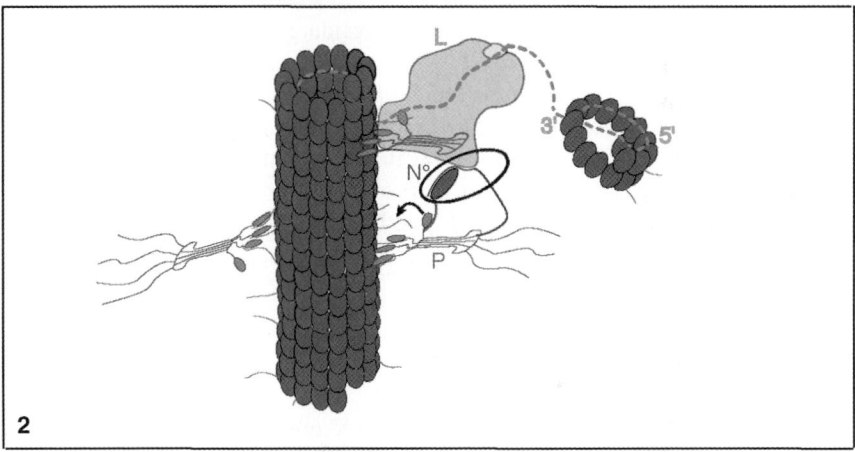

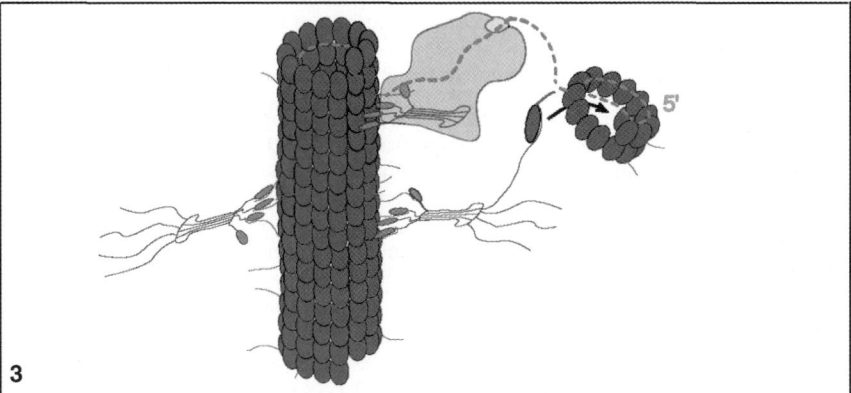

Fig. 5 Model of the polymerase complex actively replicating genomic RNA. Disordered regions are represented by *lines*. The location of the viral RNA (*dotted line*) within the N^{NUC}–P complex is schematically represented at the interior of the nucleocapsid by analogy with *Rhabdoviridae* N-RNA complexes (Albertini et al., 2006; Green et al., 2006). Within the N^{NUC}–P

Interestingly, there is a striking parallel between the N_{TAIL}–XD interaction and the PNT-N_{CORE} interaction (Fig. 5). Both interactions are not stable by themselves and must be strengthened by the combination of other interactions. This might ensure easy breaking and reforming of interactions. The relative weak affinity that typifies interacting disordered regions, together with their ability to establish contacts with other partners serving as potential regulators, would ensure dynamic breaking and reforming of interactions. This would result in transient, easily modulated interactions. One can speculate that the gain of structure of N_{TAIL} upon binding to XD could result in stabilization of the N–P complex. At the same time, N_{TAIL} folding would result in a modification in the pattern of solvent-accessible regions, resulting in the shielding of specific regions of interaction. As a result, N_{TAIL} would no longer be available for binding to its other partners. Although induced folding likely enhances the affinity between interacting proteins, the dynamic nature of these interactions could rely on (a) the intervention of viral and/or cellular cofactors modulating the strength of such interactions and (b) the ability of the IDP to establish weak affinity interactions through residual disordered regions.

Finally, since binding of N_{TAIL} to XD allows tethering of the L protein on the NC template, the N_{TAIL}–XD interaction is crucial for both viral transcription and replication. Moreover, as neither N_{TAIL} nor XD have cellular homologs, this interaction is an ideal target for antiviral inhibitors. In silico screening of small compounds for their ability to bind to the hydrophobic cleft of XD is in progress. A few candidate molecules are being tested for their ability to bind to XD and to prevent interaction with N_{TAIL} using HN-NMR (X. Morelli and S. Longhi, unpublished data).

N_{TAIL} and Molecular Partnership

The disordered nature of N_{TAIL} confers to this N domain the ability to adapt to various partners and to form complexes that are critical for both transcription and replication. Indeed, given its exposure at the surface of the viral nucleocapsid, N_{TAIL}

Fig. 5 (Continued) complex, P is represented as bound to N^{NUC} through three of its four terminal XD arms according to the model of Curran and Kolakofsky (1999). Only a few N_{TAIL} regions are drawn. PNT regions within the L–P complex have not been represented in panels 2 and 3. The P molecule delivering N° has been represented as distinct from that within the L–P complex in agreement with the results of Tuckis et al. (2002). The numbering of the different panels indicates the chronology of events. (1) L is bound to a P tetramer. A supplementary P molecule, not bound to L, is also shown (*right*). The newly synthesized RNA is shown as already partially encapsidated. (2) The encapsidation complex, N°–P, binds to the nucleocapsid template through three of its four XD arms. The extended conformation of N_{TAIL} and PNT would allow the formation of a tripartite complex between N°, P, and the polymerase (*circled*). It is tempting to imagine that the proximity of the polymerase (or an unknown signal from this) may promote the release of N° by XD, thus leading to N° incorporation within the assembling nucleocapsid. The N° release would also lead to cartwheeling of the L–P complex through binding of the free XD arm onto the nucleocapsid template (*arrow*) as in the model of Curran and Kolakofsky (1999). (3) PNT delivers N° to the newly assembled nucleocapsid (*arrow*). Reprinted with permission from Nova Publishers Inc.

establishes numerous interactions with various viral partners, including P, the P–L complex, and possibly the matrix protein (Coronel et al. 2001). Beyond viral partners, N_{TAIL} also interacts with several cellular proteins, including the heat-shock protein hsp72 (Zhang et al. 2002, 2005), the cell protein responsible for the nuclear export of N (Sato et al. 2006) and possibly components of the cell cytoskeleton (De and Banerjee 1999; Moyer et al. 1990). Moreover, N_{TAIL} within viral nucleocapsids released from infected cells also binds to the yet unidentified nucleoprotein receptor (NR) expressed at the surface of human thymic epithelial cells (Laine et al. 2003).

The interaction between N_{TAIL} and hsp72 stimulates both transcription and genome replication. Two binding sites for hsp72 have been identified (Zhang et al. 2002, 2005). High-affinity binding is supported by the α-MORE, and hsp72 can competitively inhibit binding of XD to N_{TAIL} (Zhang et al. 2005). A second low-affinity binding site is present in the C-terminus of N_{TAIL} (Zhang et al. 2002). Variability in sequence of the N protein C-terminus gives rise to hsp72 binding and nonbinding variants. Analysis of infectious virus containing a nonbinding motif shows loss of hsp72-dependent stimulation of transcription but not genome replication (Zhang et al. 2005). These findings suggest two mechanisms by which hsp72 could enhance transcription and genome replication, and both involve reducing the stability of P/N_{TAIL} complexes, thereby promoting successive cycles of binding and release that are essential to polymerase processivity (Bourhis et al. 2005; Zhang et al. 2005). The first mechanism is competition between hsp72 and XD for α-MORE binding, and this would occur at low hsp72 concentrations. In the second mechanism, hsp72 would neutralize the contribution of the C-terminus of N_{TAIL} to the formation of a stable $P-N_{TAIL}$ complex, and this would occur in the context of elevated cellular levels of hsp72 and only for MV strains that support hsp72 binding in this region (Zhang et al. 2005). The basis for the separable effects of hsp72 on genome replication versus transcription remains to be shown, with template changes unique to a replicase versus transcriptase being a primary candidate. The latter could involve unique nucleocapsid ultrastructural morphologies, with hsp72-dependent morphologies being well documented for CDV (see Oglesbee et al. 1989, 1990).

As for the functional role of hsp72 in the context of MV infection, it has been proposed that the elevation in the hsp72 levels in response to the infection could contribute to virus clearance (Carsillo et al. 2004; Oglesbee et al. 2002). Indeed, the stimulation of viral transcription and replication by hsp72 is also associated with cytopathic effects, leading to apoptosis and release of viral proteins in the extracellular compartment (Oglesbee et al. 1993; Vasconcelos et al. 1998a, 1998b). These would stimulate the adaptive immune response, thereby leading to virus clearance.

Finally, after apoptosis of infected cells, the viral nucleocapsid is released in the extracellular compartment, where it becomes available to cell surface receptors. While N_{CORE} specifically interacts with FcγRII (Laine et al. 2005), N_{TAIL} interacts with the yet uncharacterized NR, which is expressed at the surface of dendritic cells of lymphoid origin (both normal and tumoral) (Laine et al. 2003), and of T and B lymphocytes (Laine et al. 2005). The N_{TAIL}-NR interaction triggers an arrest in the G_0/G_1 phase of cell cycle, whereas the N_{CORE}–FcγRII interaction triggers apoptosis

(Laine et al. 2005). Both mechanisms have the potential to contribute to immunosuppression, which is a hallmark of MV infections (Laine et al. 2005).

Flow cytofluorimetry studies conducted on truncated forms of N_{TAIL} identified the N_{TAIL} region responsible for the interaction with NR (Box1, aa 401–420) (Laine et al. 2005). SDSL EPR studies in the presence of TFE pointed out a structural propensity within Box1 (Morin et al. 2006), the biological relevance of which may reside in a gain of structure possibly arising upon binding to NR. It is tempting to speculate that Box1 could undergo induced folding upon binding to NR. However, definitive answers on gain of regular secondary structure elements by Box1 upon binding to NR awaits the isolation of this receptor and the molecular characterization of its interaction with N_{TAIL}.

Conclusions

As thoroughly discussed in this chapter, N is a pleiotropic protein exerting multiple biological functions. The molecular bases for this pleiotropy reside in the ability of the C-terminal domain to establish interactions with various cellular and viral partners. Since many, if not all, of these interactions are critical for transcription and replication, they provide excellent targets for antiviral agents. In this context, the discovery that the N domain supporting these multiple protein interactions is intrinsically disordered is particularly relevant: indeed, protein–protein interactions mediated by disordered regions provide interesting drug discovery targets with the potential to increase the discovery rate for new compounds significantly (Cheng et al. 2006). These could eventually be used against MV and/or other *Mononegavirales* pathogens.

Acknowledgements I wish to thank all the persons who contributed to the studies described herein. In particular, within the AFMB laboratory, I would like to thank Jean-Marie Bourhis, Benjamin Morin, Stéphanie Costanzo, Sabrina Rouger, Elodie Liquière, Bruno Canard, Kenth Johansson, David Karlin, François Ferron, Véronique Receveur-Brechot, Hervé Darbon, Valérie Campanacci, and Christian Cambillau. I also thank Keith Dunker (Indiana University, USA), David Bhella (MRC, Glasgow, UK), Michael Oglesbee (Ohio State University, USA), Hélène Valentin, and Chantal Rabourdin-Combe (INSERM, Lyon, France), André Fournel, Valérie Belle, and Bruno Guigliarelli (Bioénergetique et Ingéniere des Protéins, CNRS, Marseille, France). I wish to express my gratitude to Frédéric Carrière (Enzymology at Interfaces and Physiology of Lipolysis, CNRS, Marseille, France) for having introduced me to EPR spectroscopy and for constant support. I am also grateful to David Bhella and David Karlin who are the authors of Figures 4 and 5, respectively. I also want to thank Denis Gerlier for stimulating discussions and for helpful comments on the manuscript. The studies mentioned in this chapter were conducted with the financial support of the European Commission, program RTD, QLK2-CT2001–01225, "Towards the design of new potent antiviral drugs: structure-function analysis of *Paramyxoviridae* polymerase," of the Agence Nationale de la Recherche, specific program "Microbiologie et Immunologie", ANR-05-MIIM-035–02, "Structure and disorder of measles virus nucleoprotein: molecular partnership and functional impact", and of the National Institute of Neurological Disorders and Stroke, specific program "The cellular stress response in viral encephalitis", R01 NS031693–11A2.

References

Albertini AAV, Schoehn G, Ruigrok RW (2005) Structures impliquées dans la réplication et la transcription des virus à ARN non segmentés de sens négatif. Virologie 9:83–92

Albertini AA, Wernimont AK, Muziol T, Ravelli RB, Clapier CR, Schoehn G, Weissenhorn W, Ruigrok RW (2006) Crystal structure of the rabies virus nucleoprotein-RNA complex. Science 313:360–363

Baker SC, Moyer SA (1988) Encapsidation of Sendai virus genome RNAs by purified NP protein during in vitro replication. J Virol 62:834–838

Bankamp B, Horikami SM, Thompson PD, Huber M, Billeter M, Moyer SA (1996) Domains of the measles virus N protein required for binding to P protein and self-assembly. Virology 216:272–277

Belle V, Rouger S, Constanzo S, Liquière E, Strancar J, Guigliarelli B, Fournel A, and Longhi S (2008) Mapping alpha-helical induced folding within the intrinsically disordered C-terminal domain of the measles virus nucleoprotein, by site-directed spin-labeling EPR spectroscopy. Proteins: Structure, Function and Bioinformatics (In press)

Bernado P, Blanchard L, Timmins P, Marion D, Ruigrok RW, Blackledge M (2005) A structural model for unfolded proteins from residual dipolar couplings and small-angle x-ray scattering. Proc Natl Acad Sci U S A 102:17002–17007

Bhella D, Ralph A, Murphy LB, Yeo RP (2002) Significant differences in nucleocapsid morphology within the *Paramyxoviridae*. J Gen Virol 83:1831–1839

Bhella D, Ralph A, Yeo RP (2004) Conformational flexibility in recombinant measles virus nucleocapsids visualised by cryo-negative stain electron microscopy and real-space helical reconstruction. J Mol Biol 340:319–331

Biswas R, Kuhne H, Brudvig GW, Gopalan V (2001) Use of EPR spectroscopy to study macromolecular structure and function. Sci Prog 84:45–67

Bitko V, Barik S (2001) Phenotypic silencing of cytoplasmic genes using sequence-specific double-stranded short interfering RNA and its application in the reverse genetics of wild type negative-strand RNA viruses. BMC Microbiol 1:34

Blanchard L, Tarbouriech N, Blackledge M, Timmins P, Burmeister WP, Ruigrok RW, Marion D (2004) Structure and dynamics of the nucleocapsid-binding domain of the Sendai virus phosphoprotein in solution. Virology 319:201–211

Blumberg BM, Giorgi C, Kolakofsky D (1983) N protein of vesicular stomatitis virus selectively encapsidates leader RNA in vitro. Cell 32:559–567

Bourhis J, Johansson K, Receveur-Bréchot V, Oldfield CJ, Dunker AK, Canard B, Longhi S (2004) The C-terminal domain of measles virus nucleoprotein belongs to the class of intrinsically disordered proteins that fold upon binding to their physiological partner. Virus Res 99:157–167

Bourhis JM, Canard B, Longhi S (2005a) Désordre structural au sein du complexe réplicatif du virus de la rougeole: implications fonctionnelles. Virologie 9:367–383

Bourhis JM, Receveur-Bréchot V, Oglesbee M, Zhang X, Buccellato M, Darbon H, Canard B, Finet S, Longhi S (2005b) The intrinsically disordered C-terminal domain of the measles virus nucleoprotein interacts with the C-terminal domain of the phosphoprotein via two distinct sites and remains predominantly unfolded. Protein Sci 14:1975–1992

Bourhis JM, Canard B, Longhi S (2006) Structural disorder within the replicative complex of measles virus: functional implications. Virology 344:94–110

Bourhis J, Canard B, Longhi S (2007) Predicting protein disorder and induced folding: from theoretical principles to practical applications. Curr Protein Peptide Sci 8:135–149

Buchholz CJ, Spehner D, Drillien R, Neubert WJ, Homann HE (1993) The conserved N-terminal region of Sendai virus nucleocapsid protein NP is required for nucleocapsid assembly. J Virol 67:5803–5812

Buchholz CJ, Retzler C, Homann HE, Neubert WJ (1994) The carboxy-terminal domain of Sendai virus nucleocapsid protein is involved in complex formation between phosphoprotein and nucleocapsid- like particles. Virology 204:770–776

Carsillo T, Carsillo M, Niewiesk S, Vasconcelos D, Oglesbee M (2004) Hyperthermic pre-conditioning promotes measles virus clearance from brain in a mouse model of persistent infection. Brain Res 1004:73–82

Chen JW, Romero P, Uversky VN, Dunker AK (2006a) Conservation of intrinsic disorder in protein domains and families: I. A database of conserved predicted disordered regions. J Proteome Res 5:879–887

Chen JW, Romero P, Uversky VN, Dunker AK (2006b) Conservation of intrinsic disorder in protein domains and families: II. functions of conserved disorder. J Proteome Res 5: 888–898

Chen M, Cortay JC, Gerlier D (2003) Measles virus protein interactions in yeast: new findings and caveats. Virus Res 98:123–129

Cheng Y, Legall T, Oldfield CJ, Mueller JP, Van YY, Romero P, Cortese MS, Uversky VN, Dunker AK (2006) Rational drug design via intrinsically disordered protein. Trends Biotechnol 24:435–442

Coronel EC, Takimoto T, Murti KG, Varich N, Portner A (2001) Nucleocapsid incorporation into parainfluenza virus is regulated by specific interaction with matrix protein. J Virol 75:1117–1123

Curran JA, Kolakofsky D (1991) Rescue of a Sendai virus DI genome by other parainfluenza viruses: implications for genome replication. Virology 182:168–176

Curran J, Kolakofsky D (1999) Replication of paramyxoviruses. Adv Virus Res 54:403–422

Curran J, Homann H, Buchholz C, Rochat S, Neubert W, Kolakofsky D (1993) The hypervariable C-terminal tail of the Sendai paramyxovirus nucleocapsid protein is required for template function but not for RNA encapsidation. J. Virol 67:4358–4364

Curran J, Pelet T, Kolakofsky D (1994) An acidic activation-like domain of the Sendai virus P protein is required for RNA synthesis and encapsidation. Virology 202:875–884

Curran J, Marq JB, Kolakofsky D (1995) An N-terminal domain of the Sendai paramyxovirus P protein acts as a chaperone for the NP protein during the nascent chain assembly step of genome replication. J. Virol 69:849–855

De BP, Banerjee AK (1999) Involvement of actin microfilaments in the transcription/replication of human parainfluenza virus type 3: possible role of actin in other viruses. Microsc Res Tech 47:114–123

Dunker AK, Obradovic Z (2001) The protein trinity—linking function and disorder. Nat Biotechnol 19:805–806

Dunker AK, Garner E, Guilliot S, Romero P, Albrecht K, Hart J, Obradovic Z, Kissinger C, Villafranca JE (1998) Protein disorder and the evolution of molecular recognition: theory, predictions and observations. Pac Symp Biocomput 3:473–484

Dunker AK, Lawson JD, Brown CJ, Williams RM, Romero P, Oh JS, Oldfield CJ, Campen AM, Ratliff CM, Hipps KW, Ausio J, Nissen MS, Reeves R, Kang C, Kissinger CR, Bailey RW, Griswold MD, Chiu W, Garner EC, Obradovic Z (2001) Intrinsically disordered protein. J Mol Graph Model 19:26–59

Dunker AK, Brown CJ, Lawson JD, Iakoucheva LM, Obradovic Z (2002) Intrinsic disorder and protein function. Biochemistry 41:6573–6582

Dunker AK, Cortese MS, Romero P, Iakoucheva LM, Uversky VN (2005) Flexible nets. FEBS J 272:5129–5148

Dyson HJ, Wright PE (2002) Coupling of folding and binding for unstructured proteins. Curr Opin Struct Biol 12:54–60

Dyson HJ, Wright PE (2005) Intrinsically unstructured proteins and their functions. Nat Rev Mol Cell Biol 6:197–208

Egelman EH, Wu SS, Amrein M, Portner A, Murti G (1989) The Sendai virus nucleocapsid exists in at least four different helical states. J Virol 63:2233–2243

Fearns R, Collins PL (1999) Role of the M2-1 transcription antitermination protein of respiratory syncytial virus in sequential transcription. J Virol 73:5852–5864

Feix JB, Klug CS (1998) Site-directed spin-labeling of membrane proteins and peptide-membrane interactions. In: Berliner L (ed) Biological magnetic resonance. Plenum Press, New York, pp 251–281

Ferron F, Longhi S, Henrissat B, Canard B (2002) Viral RNA-polymerases—a predicted 2'-O-ribose methyltransferase domain shared by all Mononegavirales. Trends Biochem Sci 27: 222–224

Ferron F, Longhi S, Canard B, Karlin D (2006) A practical overview of protein disorder prediction methods. Proteins 65:1–14

Fink AL (2005) Natively unfolded proteins. Curr Opin Struct Biol 15:35–41

Flamand A, Raux H, Gaudin Y, Ruigrok RW (1993) Mechanisms of rabies virus neutralization. Virology 194:302–313

Fuxreiter M, Simon I, Friedrich P, Tompa P (2004) Preformed structural elements feature in partner recognition by intrinsically unstructured proteins. J Mol Biol 338:1015–1026

Garner E, Romero P, Dunker AK, Brown C, Obradovic Z (1999) Predicting binding regions within disordered proteins. Genome Inform Ser Workshop Genome Inform 10:41–50

Giraudon P, Jacquier MF, Wild TF (1988) Antigenic analysis of African measles virus field isolates: identification and localisation of one conserved and two variable epitope sites on the NP protein. Virus Res 10:137–152

Gombart AF, Hirano A, Wong TC (1993) Conformational maturation of measles virus nucleocapsid protein. J Virol 67:4133–4141

Green TJ, Zhang X, Wertz GW, Luo M (2006) Structure of the vesicular stomatitis virus nucleoprotein-RNA complex. Science 313:357–360

Gunasekaran K, Tsai CJ, Kumar S, Zanuy D, Nussinov R (2003) Extended disordered proteins: targeting function with less scaffold. Trends Biochem Sci 28:81–85

Gupta AK, Shaji D, Banerjee AK (2003) Identification of a novel tripartite complex involved in replication of vesicular stomatitis virus genome RNA. J Virol 77:732–738

Hartlieb B, Modrof J, Muhlberger E, Klenk HD, Becker S (2003) Oligomerization of Ebola virus VP30 is essential for viral transcription and can be inhibited by a synthetic peptide. J Biol Chem 278:41830–41806

Harty RN, Palese P (1995) Measles virus phosphoprotein (P) requires the NH2- and COOH-terminal domains for interactions with the nucleoprotein (N) but only the COOH terminus for interactions with itself. J Gen Virol 76:2863–2867

Heggeness MH, Scheid A, Choppin PW (1980) Conformation of the helical nucleocapsids of paramyxoviruses and vesicular stomatitis virus: reversible coiling and uncoiling induced by changes in salt concentration. Proc Natl Acad Sci U S A 77:2631–2635

Heggeness MH, Scheid A, Choppin PW (1981) The relationship of conformational changes in the Sendai virus nucleocapsid to proteolytic cleavage of the NP polypeptide. Virology 114:555–562

Horikami SM, Moyer SA (1995) Structure, transcription, and replication of measles virus. Curr Top Microbiol Immunol 191:35–50

Hubbell WL, Gross A, Langen R, Lietzow MA (1998) Recent advances in site-directed spin labeling of proteins. Curr Opin Struct Biol 8: 649–656

Huber M, Cattaneo R, Spielhofer P, Orvell C, Norrby E, Messerli M, Perriard JC, Billeter MA (1991) Measles virus phosphoprotein retains the nucleocapsid protein in the cytoplasm. Virology 185:299–308

Iakoucheva LM, Brown CJ, Lawson JD, Obradovic Z, Dunker AK (2002) Intrinsic disorder in cell-signaling and cancer-associated proteins. J Mol Biol 323:573–584

Iseni F, Baudin F, Blondel D, Ruigrok RW (2000) Structure of the RNA inside the vesicular stomatitis virus nucleocapsid. Rna 6:270–281

Johansson K, Bourhis JM, Campanacci V, Cambillau C, Canard B, Longhi S (2003) Crystal structure of the measles virus phosphoprotein domain responsible for the induced folding of the C-terminal domain of the nucleoprotein. J Biol Chem 278:44567–44573

Karlin D, Longhi S, Canard B (2002a) Substitution of two residues in the measles virus nucleoprotein results in an impaired self-association. Virology 302:420–432

Karlin D, Longhi S, Receveur V, Canard B (2002b) The N-terminal domain of the phosphoprotein of morbilliviruses belongs to the natively unfolded class of proteins. Virology 296:251–262

Karlin D, Ferron F, Canard B, Longhi S (2003) Structural disorder and modular organization in Paramyxovirinae N and P. J Gen Virol 84:3239–3252

Kingston RL, Hamel DJ, Gay LS, Dahlquist FW, Matthews BW (2004a) Structural basis for the attachment of a paramyxoviral polymerase to its template. Proc Natl Acad Sci U S A 101:8301–8306

Kingston RL, Walter AB, Gay LS (2004b) Characterization of nucleocapsid binding by the measles and the mumps virus phosphoprotein. J Virol 78:8615–8629

Kolakofsky D, Roux L, Garcin D, Ruigrok RW (2005) Paramyxovirus mRNA editing, the "rule of six" and error catastrophe: a hypothesis. J Gen Virol 86:1869–1877

Kouznetzoff A, Buckle M, Tordo N (1998) Identification of a region of the rabies virus N protein involved in direct binding to the viral RNA. J Gen Virol 79:1005–1013

Lacy ER, Filippov I, Lewis WS, Otieno S, Xiao L, Weiss S, Hengst L, Kriwacki RW (2004) p27 binds cyclin-CDK complexes through a sequential mechanism involving binding-induced protein folding. Nat Struct Mol Biol 11:358–364

Laine D, Trescol-Biémont M, Longhi S, Libeau G, Marie J, Vidalain P, Azocar O, Diallo A, Canard B, Rabourdin-Combe C, Valentin H (2003) Measles virus nucleoprotein binds to a novel cell surface receptor distinct from FcgRII via its C-terminal domain: role in MV-induced immunosuppression. J Virol 77:11332–11346

Laine D, Bourhis J, Longhi S, Flacher M, Cassard L, Canard B, Sautès-Fridman C, Rabourdin-Combe C, Valentin H (2005) Measles virus nucleoprotein induces cell proliferation arrest and apoptosis through NTAIL/NR and NCORE/FcγRIIB1 interactions, respectively. J Gen Virol 86:1771–1784

Lamb RA, Kolakofsky D (2001) *Paramyxoviridae* : the viruses and their replication. In: Fields BN, Knipe DM, Howley PM (eds) Fields virology, 4th edn., Lippincott-Raven, Philadelphia, pp 1305–1340

Li J, Wang JT, Whelan SP (2006) A unique strategy for mRNA cap methylation used by vesicular stomatitis virus. Proc Natl Acad Sci U S A 103:8493–8498

Lieutaud P, Canard B, Longhi S (2008) MeDor: a metaserver for predicting protein disorder. BMC Genomics 9(Suppl 2):S25

Liston P, Batal R, DiFlumeri C, Briedis DJ (1997) Protein interaction domains of the measles virus nucleocapsid protein (NP). Arch Virol 142:305–321

Longhi S, Canard B (1999) Mécanismes de transcription et de réplication des *Paramyxoviridae*. Virologie 3:227–240

Longhi S, Receveur-Brechot V, Karlin D, Johansson K, Darbon H, Bhella D, Yeo R, Finet S, Canard B (2003) The C-terminal domain of the measles virus nucleoprotein is intrinsically disordered and folds upon binding to the C-terminal moiety of the phosphoprotein. J Biol Chem 278:18638–18648

Luo M, Green TJ, Zhang X, Tsao J, Qiu S (2007) Conserved characteristics of the rhabdovirus nucleoprotein. Virus Res 129:246–251

Marion D, Tarbouriech N, Ruigrok RW, Burmeister WP, Blanchard L (2001) Assignment of the 1H, 15N and 13C resonances of the nucleocapsid- binding domain of the Sendai virus phosphoprotein. J Biomol NMR 21:75–76

Mavrakis M, Kolesnikova L, Schoehn G, Becker S, Ruigrok RW (2002) Morphology of Marburg virus NP-RNA. Virology 296:300–307

Mohan A, Oldfield CJ, Radivojac P, Vacic V, Cortese MS, Dunker AK, Uversky VN (2006) Analysis of molecular recognition features (MoRFs). J Mol Biol 362:1043–1059

Morin B, Bourhis JM, Belle V, Woudstra M, Carrière F, BGuigliarelli B, Fournel A, Longhi S (2006) Assessing induced folding of an intrinsically disordered protein by site-directed spin-labeling EPR spectroscopy. J Phys Chem B 110(41) 20596–20608

Moyer SA, Baker SC, Horikami SM (1990) Host cell proteins required for measles virus reproduction. J Gen Virol 71:775–783

Myers TM, Pieters A, Moyer SA (1997) A highly conserved region of the Sendai virus nucleocapsid protein contributes to the NP-NP binding domain. Virology 229:322–335

Myers TM, Smallwood S, Moyer SA (1999) Identification of nucleocapsid protein residues required for Sendai virus nucleocapsid formation and genome replication. J Gen Virol 80:1383–1391

Ogino T, Kobayashi M, Iwama M, Mizumoto K (2005) Sendai virus RNA-dependent RNA polymerase L protein catalyzes cap methylation of virus-specific mRNA. J Biol Chem 280:4429–4435

Oglesbee M, Ringler S, Krakowka S (1990) Interaction of canine distemper virus nucleocapsid variants with 70 K heat-shock proteins. J Gen Virol 71:1585–1590

Oglesbee M, Tatalick L, Rice J, Krakowka S (1989) Isolation and characterization of canine distemper virus nucleocapsid variants. J Gen Virol 70:2409–2419

Oglesbee MJ, Kenney H, Kenney T, Krakowka S (1993) Enhanced production of morbillivirus gene-specific RNAs following induction of the cellular stress response in stable persistent infection. Virology 192:556–567

Oglesbee MJ, Pratt M, Carsillo T (2002) Role for heat shock proteins in the immune response to measles virus infection. Viral Immunol 15:399–416

Oldfield CJ, Cheng Y, Cortese MS, Romero P, Uversky VN, Dunker AK (2005) Coupled folding and binding with alpha-helix-forming molecular recognition elements. Biochemistry 44:12454–12470

Plumet S, Duprex WP, Gerlier D (2005) Dynamics of viral RNA synthesis during measles virus infection. J Virol 79:6900–6908

Rahaman A, Srinivasan N, Shamala N, Shaila MS (2004) Phosphoprotein of the rinderpest virus forms a tetramer through a coiled coil region important for biological function. A structural insight. J Biol Chem 279:23606–23614

Receveur-Bréchot V, Bourhis JM, Uversky VN, Canard B, Longhi S (2006) Assessing protein disorder and induced folding. Proteins 62:24–45

Robbins SJ, Bussell RH (1979) Structural phosphoproteins associated with purified measles virions and cytoplasmic nucleocapsids. Intervirology 12:96–102

Robbins SJ, Bussell RH, Rapp F (1980) Isolation and partial characterization of two forms of cytoplasmic nucleocapsids from measles virus-infected cells. J Gen Virol 47:301–310

Roux L (2005) Dans le génome des Paramyxovirinae, les promoteurs et leurs activités sont façonnés par la "règle de six". Virologie 9:19–34

Ryan KW, Portner A (1990) Separate domains of Sendai virus P protein are required for binding to viral nucleocapsids. Virology 174:515–521

Sato H, Masuda M, Miura R, Yoneda M, Kai C (2006) Morbillivirus nucleoprotein possesses a novel nuclear localization signal and a CRM1-independent nuclear export signal. Virology 352:121–130

Schoehn G, Iseni F, Mavrakis M, Blondel D, Ruigrok RW (2001) Structure of recombinant rabies virus nucleoprotein-RNA complex and identification of the phosphoprotein binding site. J Virol 75:490–498

Schoehn G, Mavrakis M, Albertini A, Wade R, Hoenger A, Ruigrok RW (2004) The 12 A structure of trypsin-treated measles virus N-RNA. J Mol Biol 339:301–312

Shoemaker BA, Portman JJ, Wolynes PG (2000) Speeding molecular recognition by using the folding funnel: the fly-casting mechanism. Proc Natl Acad Sci U S A 97(16) 8868–8873

Sivakolundu SG, Bashford D, Kriwacki RW (2005) Disordered p27Kip1 exhibits intrinsic structure resembling the Cdk2/cyclin A-bound conformation. J Mol Biol 353:1118–1128

Smallwood S, Ryan KW, Moyer SA (1994) Deletion analysis defines a carboxyl-proximal region of Sendai virus P protein that binds to the polymerase L protein. Virology 202:154–163

Spehner D, Kirn A, Drillien R (1991) Assembly of nucleocapsid-like structures in animal cells infected with a vaccinia virus recombinant encoding the measles virus nucleoprotein. J Virol 65:6296–6300

Spehner D, Drillien R, Howley PM (1997) The assembly of the measles virus nucleoprotein into nucleocapsid-like particles is modulated by the phosphoprotein. Virology 232:260–268

Stallcup KC, Wechsler SL, Fields BN (1979) Purification of measles virus and characterization of subviral components. J Virol 30:166–176

Sugase K, Dyson HJ, Wright PE (2007) Mechanism of coupled folding and binding of an intrinsically disordered protein. Nature 447:1021–1025

Sweetman DA, Miskin J, Baron MD (2001) Rinderpest virus C and V proteins interact with the major (L) component of the viral polymerase. Virology 281:193–204

Tapparel C, Maurice D, Roux L (1998) The activity of Sendai virus genomic and antigenomic promoters requires a second element past the leader template regions: a motif (GNNNNN)3 is essential for replication. J Virol 72:3117–3128

Tarbouriech N, Curran J, Ruigrok RW, Burmeister WP (2000) Tetrameric coiled coil domain of Sendai virus phosphoprotein. Nat Struct Biol 7:777–781

Tompa P (2002) Intrinsically unstructured proteins. Trends Biochem Sci 27:527–533

Tompa P (2003) The functional benefits of disorder. J Mol Structure (Theochem) 666–667:361–371

Tompa P (2005) The interplay between structure and function in intrinsically unstructured proteins. FEBS Lett 579:3346–3354

Tran TL, Castagne N, Bhella D, Varela PF, Bernard J, Chilmonczyk S, Berkenkamp S, Benhamo V, Grznarova K, Grosclaude J, Nespoulos C, Rey FA, Eleouet JF (2007) The nine C-terminal amino acids of the respiratory syncytial virus protein P are necessary and sufficient for binding to ribonucleoprotein complexes in which six ribonucleotides are contacted per N protein protomer. J Gen Virol 88:196–206

Tsai CD, Ma B, Kumar S, Wolfson H, Nussinov R (2001a) Protein folding: binding of conformationally fluctuating building blocks via population selection. Crit Rev Biochem Mol Biol 36:399–433

Tsai CJ, Ma B, Sham YY, Kumar S, Nussinov R (2001b) Structured disorder and conformational selection. Proteins 44:418–427

Tuckis J, Smallwood S, Feller JA, Moyer SA (2002) The C-terminal 88 amino acids of the Sendai virus P protein have multiple functions separable by mutation. J Virol 76:68–77

Uversky VN (2002a) Natively unfolded proteins: a point where biology waits for physics. Protein Sci 11:739–756

Uversky VN (2002b) What does it mean to be natively unfolded? Eur J Biochem 269:2–12

Uversky VN, Li J, Souillac P, Jakes R, Goedert M, Fink AL (2002) Biophysical properties of the synucleins and their propensities to fibrillate: inhibition of alpha-synuclein assembly by beta- and gamma-synucleins. J Biol Chem 277:11970–11978

Uversky VN, Oldfield CJ, Dunker AK (2005) Showing your ID: intrinsic disorder as an ID for recognition, regulation and cell signaling. J Mol Recognit 18:343–384

Vacic V, Oldfield CJ, Mohan A, Radivojac P, Cortese MS, Uversky VN, Dunker AK (2007) Characterization of molecular recognition features, MoRFs, and their binding partners. J Proteome Res 6:2351–2366

Vasconcelos DY, Cai XH, Oglesbee MJ (1998a) Constitutive overexpression of the major inducible 70 kDa heat shock protein mediates large plaque formation by measles virus. J Gen Virol 79:2239–2247

Vasconcelos D, Norrby E, Oglesbee M (1998b) The cellular stress response increases measles virus-induced cytopathic effect. J Gen Virol 79:1769–1773

Vincent S, Tigaud I, Schneider H, Buchholz CJ, Yanagi Y, Gerlier D (2002) Restriction of measles virus RNA synthesis by a mouse host cell line: trans-complementation by polymerase components or a human cellular factor(s). J Virol 76:6121–6130

Vulliemoz D, Roux L (2001) "Rule of six": how does the Sendai virus RNA polymerase keep count? J Virol 75:4506–4518

Warnes A, Fooks AR, Dowsett AB, Wilkinson GW, Stephenson JR (1995) Expression of the measles virus nucleoprotein gene in *Escherichia coli* and assembly of nucleocapsid-like structures. Gene 160:173–178

Wright PE, Dyson HJ (1999) Intrinsically unstructured proteins: re-assessing the protein structure-function paradigm. J Mol Biol 293:321–331

Zhang X, Glendening C, Linke H, Parks CL, Brooks C, Udem SA, Oglesbee M (2002) Identification and characterization of a regulatory domain on the carboxyl terminus of the measles virus nucleocapsid protein. J Virol 76:8737–8746

Zhang X, Bourhis JM, Longhi S, Carsillo T, Buccellato M, Morin B, Canard B, Oglesbee M (2005) Hsp72 recognizes a P binding motif in the measles virus N protein C-terminus. Virology 337:162–174

Chapter 7
Reverse Genetics of Measles Virus and Resulting Multivalent Recombinant Vaccines: Applications of Recombinant Measles Viruses

M.A. Billeter(✉), H.Y. Naim, and S.A. Udem

Contents

Development of Techniques Allowing Reverse Genetics of RNA Viruses	130
Positive-Strand RNA Viruses	131
Negative-Strand RNA Viruses	132
Genomic Modifications of Existing Measles Virus Strains	138
Measles Viruses with Inserted Foreign Coding Sequences	140
Technical Aspects of Foreign Gene Expression	140
Insertion of Genes Encoding Markers and Immunomodulators	142
Insertion of Genes from Other Pathogens: Candidate Multivalent Vaccines	143
Creation of Targeted MVs with Oncolytic Properties	147
Development of Segmented MVs Carrying Foreign Genes	147
Considerations for Practical Applications of Recombinant Multivalent MV Vaccines	148
Stability of Recombinant MVs	148
Efficacy and Safety	149
Preimmunity: Prime-Boost Regimens	151
Impediments Imposed by Regulatory and Commercial Issues	152
Other Recombinant *Mononegavirales* as Vaccines	154
Conclusions	155
References	156

Abstract An overview is given on the development of technologies to allow reverse genetics of RNA viruses, i.e., the rescue of viruses from cDNA, with emphasis on nonsegmented negative-strand RNA viruses (*Mononegavirales*), as exemplified for measles virus (MV). Primarily, these technologies allowed site-directed mutagenesis, enabling important insights into a variety of aspects of the biology of these viruses. Concomitantly, foreign coding sequences were inserted to (a) allow localization of virus replication in vivo through marker gene expression, (b) develop candidate multivalent vaccines against measles and other pathogens, and (c) create candidate oncolytic viruses. The vector use of these viruses was experimentally encouraged by the pronounced genetic stability of the recombinants unexpected for RNA viruses, and by the high load

M.A. Billeter
University of Zurich, Winterthurerstrasse 190, 8057 Zurich, Switzerland, e-mail: billeter@access.unizh.ch

of insertable genetic material, in excess of 6 kb. The known assets, such as the small genome size of the vector in comparison to DNA viruses proposed as vectors, the extensive clinical experience of attenuated MV as vaccine with a proven record of high safety and efficacy, and the low production cost per vaccination dose are thus favorably complemented.

Development of Techniques Allowing Reverse Genetics of RNA Viruses

Reverse genetics, i.e., the introduction of exactly defined mutations into the genome of organisms (true living organisms and viruses) has become one of the technical cornerstones of molecular biology. Convenient application of this important technique for RNA viruses relies on the possibility to rescue such viruses from cloned DNA that represent sequence variants of their RNA genomes. The underlying DNA segments can be easily obtained, basically by reverse transcription in vitro, manipulated, even conveniently synthesized chemically, and economically amplified in bacteria as segments residing within plasmids.

The primary driving force for establishing reverse genetic systems for RNA viruses was to obtain insights into their biology. Indeed, investigations using this tool are bringing about a wealth of knowledge regarding the propagation of viruses and their interactions with their hosts. In addition, it was recognized early on that reverse genetic technology might also be exploited for practical purposes in medicine, to create new vaccines, and possibly also to combat cancer, taking advantage of the strong and often long-lasting stimulation of immune systems by viruses. Particularly promising was the use of attenuated viruses already widely proven in medical practice to constitute efficient and safe vaccines against their parent wild-type progenitors. Addition to their genome of genetic material encoding proteins or protein fragments of other important pathogens might result in multivalent vaccines mediating protection not only against the virus species used as vector, but also against the pathogens from which the added genetic information is derived.

Over the three decades since the 1970s, great progress has been made to gradually make essentially all RNA virus genera amenable to the application of reverse genetics, and during the last decade the processes to obtain genetically modified viruses by reverse genetics have been refined to facilitate the technical processes, so that practically any desired constructed RNA virus can be made readily available, provided that the changes introduced are compatible with efficient propagation of the construct in cultured cells and packaging into infectious viral particles. As outlined below, manipulated plus-strand RNA viruses were obtained rather easily, and among the more problematic negative-strand RNA viruses, those containing a nonsegmented genome (i.e., the ones grouped in the large order *Mononegavirales*) were more difficult to get hold of than representatives containing segmented genomes.

Positive-Strand RNA Viruses

The bacteriophage Qbeta was the first virus for which this technique could be applied (Taniguchi et al. 1978). In this case, the rescue of virus was particularly easy, as the cloned DNA segment representing the sequence of the viral genome, simply flanked by oligo A and T stretches, was transcribed spontaneously after introduction into *Escherichia coli* cells by the residing RNA polymerase. Since the intracellularly transcribed genomes are first active as messenger RNA to produce all viral proteins and subsequently as template for the viral RNA polymerase/replicase to produce antigenomes, from which progeny genomes are then transcribed, progeny virus is formed automatically. It is still not clear by which mechanisms the primary transcripts containing additional nonviral flanking sequences give rise to progeny RNA containing the precise viral ends.

The next virus made amenable to reverse genetics was poliomyelitis virus (Racaniello and Baltimore 1981). The experimental hurdles to overcome were considerably higher in this case: (a) culturing the host cells was much more demanding, (b) the genome to be cloned was approximately twice as long, and (c) it was necessary to provide a promoter for residing DNA-dependent RNA polymerase. Nevertheless, due to some key properties similar to those outlined for Qbeta, progeny virus could be rescued spontaneously after introduction either of the properly constructed DNA or of in vitro transcribed RNA into host cells. This pioneering work triggered the rescue of a very large variety of plus-strand RNA viruses infecting vertebrates, invertebrates, plants, and microorganisms. Even coronaviruses, the largest RNA viruses with genomes in excess of 30 kb are now amenable to genetic manipulation. Very similar experimental techniques could be employed for all positive-strand RNA viruses, despite their vastly differing propagation mechanisms.

Attenuated poliomyelitis virus was also the first RNA virus to be employed as a vector to create plurivalent vaccine candidates by introduction of foreign genetic material (Andino et al. 1994). Vector use of most positive-strand RNA viruses suffers, however, from two major drawbacks. On one hand, the size of the introduced coding sequences is limited because of the space constraints in icosahedral viral capsids. On the other hand, the additional sequences introduced often survive in the original form only for a few viral generations. This was particularly evident for Qbeta and other RNA bacteriophages, which eliminate and/or modify added or altered sequences at an incredibly fast rate. The underlying reason is the strict requirement of suitable secondary and tertiary RNA structure for optimal RNA replication by the notoriously error-prone RNA-dependent RNA polymerases, usually defined as replicases. RNA structure is of paramount importance in all plus-strand RNA viruses, the genomes of which in the active state inside their host cells are naked or only loosely associated with proteins. In particular, deletions are introduced in their transcripts by recombination due to copy-choice transcription, which comes about by detachment of the replicase associated with the incomplete transcript from the template and reattachment to the template, preferentially at a location complementary to the last nucleotides of the transcript. Since RNA viruses with deletions of nonessential genome regions replicate more rapidly, such spontaneous mutants soon overgrow the original constructed genomes.

Despite these drawbacks, the genetic stability of some recombinant viruses appears sufficient for practical applications. In particular, alphavirus and flavivirus vectors are promising and have in part advanced to clinical trials. Among the former, Semliki Forest, Sindbis, and Venezuelan Equine Encephalitis viruses can be mentioned (Atkins et al. 1996; Tubulekas et al. 1997; Polo et al. 1999; Pushko et al. 1997). Among the latter, yellow fever virus based on the licensed attenuated vaccine strain (Arroyo et al. 2004; Monath et al. 2003; Brandler et al. 2005) seems particularly promising.

Negative-Strand RNA Viruses

Two distinctive features of all negative-strand RNA viruses render them inherently more difficult to rescue from cloned DNA:

1. Their genomes, which are complementary to sense strands, are not active as messenger RNA to produce the viral proteins, but are merely functional as templates required for the formation initially of mRNAs (transcription step) and subsequently of antigenomic RNA (replication step 1), which then acts as template for the synthesis of genomic RNA (replication step 2). Both transcription and replication are mediated by the viral RNA-dependent RNA polymerase, also defined as transcriptase/replicase. The transcription step is characterized by the formation of less than full-length transcripts, which are typically both capped and polyadenylated, whereas in the replication steps unmodified full-length transcripts are synthesized.
2. The genomic RNA is only biologically active when it is present as ribonucleoprotein (RNP), i.e., associated with at least a nucleocapsid protein (called N or NP) and the viral RNA polymerase. In other words, the minimal infectious unit consists not merely of RNA, but of a specifically associated complex, the RNP. In the case of highly segmented viruses such as Influenza viruses, the polymerase is composed of three polypeptides. In *Mononegavirales*, it consists of only one very large polypeptide (L), frequently associated with a phosphoprotein (P), which fulfills multiple functions in addition to its role as polymerase cofactor, in particular as a chaperone mediating the attachment of N to genome or antigenome RNA as they are being synthesized. Thus, for rescue of virus from cloned DNA, it is necessary to deliver RNP to the host cells by one of two approaches. In vitro-made transcripts must be functionally covered in vitro by the viral proteins mentioned above before introduction into host cells. Alternatively, transcripts, on one hand, and encapsidating proteins (in large quantities) and viral RNA polymerase (in tiny amounts), on the other hand, have to be concomitantly synthesized in the host cells to form biologically active RNPs.

Negative-Strand RNA Viruses with Highly Segmented Genomes

Almost two decades ago, the first partial rescue of a negative-strand RNA virus from cloned DNA was described (Luytjes et al. 1989; Enami et al. 1990). Since the RNP of segmented negative-strand RNA viruses is structured relatively loosely, artificial in vitro transcripts of one cloned segment of an Influenza virus strain X could be encapsidated in vitro by the proteins required for RNP formation (nucleocapsid and the three large polypeptides forming together the viral RNA polymerase). When this in vitro-formed RNP was introduced into cells infected with Influenza virus strain Y used as helper virus, reassortants of Influenza virus could be isolated containing seven segments of the helper virus Y and one segment derived from strain X. However, a method was required to separate the small number of reassortants away from the large excess of nonreassorted helper virus. This was achievable by means of specific antibodies directed against the protein encoded by the segment of strain Y to be replaced.

Although this work not only constituted the first breakthrough in this area, but also permitted the generation of some interesting designed Influenza virus mutants, experimentation was limited, since the necessity of a counter-selection step restricted the range of rescuable manipulated constructs. One decade of intensive work by several groups was required to elaborate a method enabling the rescue of Influenza virus entirely from DNA, i.e., without a helper virus (Neumann et al. 1999; Fodor et al. 1999). This method, using a large array of simultaneously co-transfected plasmids, specifying the eight antigenomic RNAs as well as the mRNAs for production of the proteins required for RNP formation, allows rescue of practically any desired Influenza virus compatible with the required biological properties of this class of viruses.

Negative-Strand RNA Viruses with Nonsegmented Genomes

Preparatory Work Using Minireplicons

Numerous research groups attempted, but invariably failed, to assemble in vitro RNPs of *Mononegavirales*, following or modifying the lessons provided in the initial accounts describing the rescue of Influenza virus. There are three main reasons for this failure: first and foremost, the RNPs of *Mononegavirales* are much tighter and more rigid than those of segmented Influenza virus (indeed, in contrast to the latter they are completely resistant to RNase digestion). Second, the RNA genomes to be covered are much longer than the one Influenza segment being encapsidated by the original Influenza virus rescue procedure. Third, the extremely large and multidomain RNA polymerase (L) of *Mononegavirales* is more prone to nonfunctional aggregation than the three subunits of the Influenza enzyme. Consequently, many investigators reasoned that genomes or antigenomes of *Mononegavirales* should be encapsidated inside cultured cells in statu nascendi, i.e., while they are being synthesized artificially from a cDNA template by a foreign DNA-dependent RNA polymerase expressed in the host cells. Indeed, soon after the successful rescue of Influenza viruses, the first steps toward rescue of *Mononegavirales* were achieved: RNA minireplicons of certain members of this viral order triggered from cloned DNA were propagated in various cell lines. The DNAs underlying these minireplicons were initially derived by reverse transcription from natural defective interfering particles (DIs) generated spontaneously when *Mononegavirales* are propagated in serial succession at high infection rates. Later, minireplicon DNAs were constructed that minimally contained the noncoding 5'- and 3'-terminal signal sequences essential for virus-specific RNA replication; the viral genes were either partially or completely deleted and replaced by reporter genes for convenient detection of virus-specific transcription and replication (Collins et al. 1991, 1993; Park et al. 1991; Conzelmann and Schnell 1994; Sidhu et al. 1995). Investigators used two kinds of helper viruses to provide the proteins to intracellularly encapsidate and replicate these minireplicons. In a simple approach, in vitro transcripts from the small cloned DNA segments were introduced into cells infected with the cognant parent nondefective virus. In a more advanced strategy, the DNA specifying the minireplicons were transfected, along with DNA encoding the encapsidating proteins nucleocapsid (N), phosphoprotein (P), and the large polymerase (L) required for formation of biologically active RNP, into cells infected with a Vaccinia virus construct (vTF7–3), engineered to express T7 RNA polymerase at high levels. All DNAs contained T7 promoters; thus, this strategy allowed efficient transcription of all virus-specific cloned DNA segments in the transfected cells.

First Rescues Employing Helper Viruses

Obviously, the goal of these experiments was to find optimal conditions to achieve replication of the much longer genomes of the full-length nondefective viruses, and then to isolate spurious amounts of assembled rescued viruses away from the helper viruses present in large quantities. This was first achieved for rabies virus (RV) (Schnell et al. 1994). Using almost identical technical approaches, approximately 1 year later, two groups were able to rescue vesicular stomatitis virus (VSV) (Lawson et al. 1995; Whelan et al. 1995). Both RV and VSV belong to the family of *Rhabdoviridae*, the members of which are characterized by a sturdy bullet-shaped virion structure, and by highly efficient, vigorous replication achieving high viral titers. Furthermore, their replication/propagation is only scarcely inhibited by the vaccinia-T7 helper virus. Thus, the tiny amounts of rescued virus could be isolated by a series of elaborate purification steps, including passage through filters.

Rescues Relying Exclusively on Transfected DNA

Although the pioneering work described in the previous section established that functional RNPs of entire genomes of *Mononegavirales* can indeed be formed intracellularly, this approach necessarily had to fail in the rescue of measles virus (MV), a representative of the family *Paramyxoviridae*, which contains larger genomes, replicates to titer orders of a magnitude lower than RV and VSV, and forms large polyploid virions that are easily inactivated by chemical and physical stress such as filtration. Therefore, it appeared desirable to develop a rescue method avoiding helper viruses altogether, thus obviating the need for a purification step. This was achieved by relying only on transfected plasmids encoding N and P proteins as well as T7 RNA polymerase; transcription of these artificial mRNAs was governed by strong CMV promoters recognized by cellular RNA polymerase II residing in nuclei, rather than by T7 promoters. An intron and a polyadenylation sequence in the transcribed region of the plasmids was inserted to ensure capping, splicing, and polyadenylation, thus enabling efficient export to the cytoplasm as well as a long half-life of the mRNAs. It was clear that, despite these precautions, expression from a plasmid would allow a much weaker expression of T7 RNA polymerase than that achievable with vaccinia-T7. However, in the strategy chosen, the plasmids encoding the viral proteins N and P, which are both required in large quantities, did not rely on T7 promoters, as in the procedure enabling rescue of *Rhabdoviridae*. Thus, a relatively low level of T7 RNA polymerase was thought to be sufficient for the synthesis of antigenomic RNA and the mRNA encoding the large viral polymerase, which is required only in tiny amounts. It should be mentioned that, to maximize the chances of success, initially an approach involving two separate steps was employed. First, helper cells were stably transfected with the three plasmids endowed with CMV promoters encoding all three proteins: N, P, and T7 RNA polymerase. The plasmid encoding T7 RNA polymerase contained a neomycin (Neo) resistance marker, and among approximately 100 neoresistant colonies, two were chosen to propagate lines of helper cells, which, in addition to T7 RNA polymerase, express both N and P at appreciable levels. (These helper cell lines are still being used today by several research groups). Second, the helper cells were transiently transfected with two plasmids containing T7 promoters. One specified precisely initiated and terminated minireplicon (and later full-length antigenomic) transcripts; the other, used in very small amounts, encoded the large viral RNA polymerase L.

This approach triggered high levels of chloramphenicol acetyl transferase (CAT) in cells transfected with artificial minireplicon DNA bearing a CAT reporter gene, similar to the level that had been obtained by introducing the in vitro transcripts of CAT minireplicon DNA into MV-infected cells. In further experiments, successively longer viral antigenomic regions were added to the minireplicon sequence, giving rise to what is termed midireplicons. As can be expected, the CAT signal elicited became weaker with increasing midireplicon size. Finally, all of the missing antigenomic sequence was added, leading, at the end of 1994, to the first rescued MVs, which were clonally derived from single syncytia, thus constituting an additional benefit in comparison to rescues using a helper virus that yields mixtures of different clones (see Fig. 1A). This virus was denominated MV Edtag, as it had been tagged by three nucleotide exchanges in a nontranslated region, to distinguish it from the parent attenuated vaccine-derived MV Edmonston B type strain passaged extensively in many laboratories. However, it took more than 6 months before

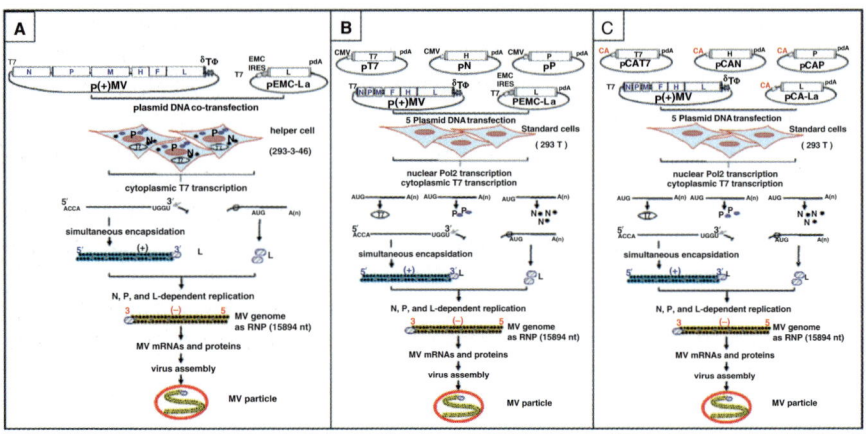

Fig. 1 A–C Rescue methods for MV strains and recombinant MVs. **A** Rescue as first described (Radecke et al. 1995), using as helper cells a clone of 293 cells (293–3-46) stably transfected with expression plasmids encoding T7 RNA polymerase, N and P proteins. Note that the intracellular T7 antigenomic transcripts specified by the p(+)MV type plasmids start precisely at the antigenomic 5' end and are cleaved automatically by the hepatitis ' ribozyme sequence after termination at the T7 termination signal T'. **B** Slightly modified rescue protocol: simultaneous transfection of 293 cells or any other cell line supporting replication of MV, using exactly the same five plasmids used in the original method, where the plasmids pT7, pN, and pP had been utilized to generate the 293–3-46 cell clone. **C** Similar method as in **B**, but using the highly efficient chicken actin (CA) promoter rather than the cytomegalovirus (CMV) promoter (Martin et al. 2006)

a patent application was filed, and almost 1 year for the first original research paper to be published (Radecke et al. 1995), since all experimental procedures had to be verified by different collaborators, and the genetic stability of rescued viruses had to be ascertained by adding marker genes. Even chimera were constructed where the envelope proteins of MV were genetically substituted by the unique envelope protein of another virus, VSV (Spielhofer 1995; Spielhofer et al. 1998).

A comment regarding all rescue procedures in general is necessary here. It might appear odd to artificially express antigenomic rather than genomic RNA, since this requires a replication step to yield genomic RNP, from which transcription to produce all viral mRNAs and subsequently all viral proteins can take place. However, several research groups independently adopted this strategy, which circumvents hybridization and thus inactivation of artificially expressed antigenomes with the large amounts of artificially expressed mRNA encoding N and P.

Methodological Refinements Allowing Experimentally Straightforward Rescue of Virtually All *Mononegavirales*

Soon after the publication of RV rescue, the so-called MVA-T7 virus was constructed, based on the modified virus Ankara (MVA) strain of vaccinia virus. This recombinant expresses T7 RNA polymerase at levels similar to those produced by the previously available helper virus vTF7-3. Importantly, however, the MVA variant replicates only up to a certain point in mammalian cells, and forms progeny virus only in avian cells. Thus, use of this defective helper virus, circumventing the elimination of progeny helper virus, was a second means for the rescue of the large majority of constructed *Mononegavirales* that are not obtainable by the original procedure enabling the rescue of *Rhabdoviridae*. Indeed, shortly after publication of the independent methodology described above, the rescue of human respiratory syncytial virus (RSV) employing MVA-T7 was reported (Collins et al. 1995). Numerous publications describing the rescue of different genera of *Mononegavirales* with this method, which is actually a variant of the rhabdovirus rescue procedure, appeared after this account. It should be mentioned that a further complex method only loosely similar to the RV and VSV rescue procedure, employing vTF7-3, but a completely different purification approach, applicable only for rescue of Sendai virus (SeV) and a few particular representatives of *Paramyxoviridae*, was published (Garcin et al. 1995) concurrently with the first rescue of RSV (Collins et al. 1995).

It is worth noting however, that all methods based on vaccinia-derived helper virus, regardless of whether these form progeny during the rescue procedure, suffer from several important drawbacks, the most important being the high rate of recombination induced by vaccinia virus among the transfected plasmids, which leads to frequent incorporation of sequences residing in the plasmids encoding N and P protein into the antigenomic plasmid and thus into the rescued virus. Therefore, several of the most proficient investigators in the field later switched to procedures similar to the one described above, exclusively using expression from plasmids, thus circumventing the need for a helper virus. The first published account in this line was used for the rescue of bovine RSV (Buchholz et al. 1999).

In this context, it should also be mentioned that, curiously even in 2001, Yusuke Yanagi's group described a protocol for rescuing wild-type MV strains (Tatsuo et al. 2001), which is based on the replication proficient vaccinia-T7 strain vTF7-3 also used originally for the rescue of RV. However, it should be noted that the Yanagi group used a very particular cell line shown earlier to be suitable for isolation of wild-type MV strains from acutely infected individuals, the marmoset B cell derived line B95a expressing SLAM, the receptor exclusively used by most wild-type MV strains. In this cell line, the helper virus vTF7-3 replicates exceedingly slowly and forms only a small number of progeny. Thus, the experimental condition closely resembles that using the replication-deficient MVA-T7 helper virus as discussed in the preceding paragraph.

Advantageous modifications of the original method for MV rescue involved a gradual shift from the use of persistently transfected helper cells to only transiently transfected cell lines. The modifications are particularly advantageous when new viral species are to be rescued from cloned cDNA, since they circumvent the tedious selection of cellular clones expressing, in addition to T7 RNA, the N and P proteins of the new viral species at suitable ratios. Standard cell lines able to support replication of the new virus are transfected simultaneously with all five required plasmids. (See Fig. 1B. Note that for MV rescue, the exact original transient transfection procedure is utilized, varied only by inclusion of the three plasmids expressing T7 RNA polymerase and the viral proteins N and P, which in the original set-up are already present in the persistently transfected helper cells.) The rescue efficiency using only one-step transient transfection is higher than in the original method and is enhanced further when instead of CMV promoters chicken actin promoters (Martin et al. 2006) are used (compare Fig. 1B and C). This principle was first adopted for the rescue of mumps virus (Clarke et al. 2000) and later for many manipulated representatives of *Mononegavirales*, as described in great experimental detail in a recent international patent application (WO 2004/113517 A2).

An additional simplification of MV rescue has recently been described utilizing a CMV promoter for the antigenomic expression plasmid, recognized by the nuclear polymerase II, rather than the traditional T7 promoter (Martin et al. 2006). The reason for trying this strategy was actually dictated by Borna virus (BV) rescue. BV replicates in the nucleus; therefore, it was reasonable to test pol I and pol II promoters, and the latter proved to be optimal. The finding that this strategy was also successful for MV suggests that MV RNP can be formed in the nucleus concomitantly with the synthesis of antigenomic RNA by pol II, thus impeding splicing at fortuitous splice sites, and that RNPs then reach the cytoplasm rather efficiently. RNP formation in the nucleus appears likely attributable to the fact that N protein, when artificially expressed alone, migrates efficiently to the nucleus (Huber et al. 1991) and likely also carries along some associated chaperone P when both proteins are simultaneously expressed.

Genomic Modifications of Existing Measles Virus Strains

As mentioned above, there was one major driving force behind the development of systems enabling reverse genetics for *Mononegavirales*: to gain better insight into the biology of the representatives of this viral order, as more comprehensively illustrated for MV in other chapters of this book. Since this chapter mainly deals with the possible practical applications enabled by these techniques, only some of the first designed artificial mutants of MV are mentioned here.

The first of these modifications consisted in a large deletion of 504 nucleotides, almost the entire 5' untranslated region (UTR) of the F gene (Radecke et al. 1995), which is very GC-rich. In cell cultures, this mutant did not show any deterioration in propagation. No experiments in the animal models described in other chapters (ferrets, cotton rats, transgenic mice, and macaques) have been conducted to date, which might demonstrate that the deleted region is indeed dispensable for viability in infected model animals, and thus likely also in humans. (It should be noted, however, that in human thymus/liver implants engrafted in SCID mice, this mutant replicated somewhat more slowly and the virus titer reached was roughly ten times lower than that of the parent MV Edtag; Valsamakis et al. 1998.) Also, in the F gene of other representatives of the genus morbillivirus, relatively long 5' UTRs exist, although none with a comparable length. Recently, this region as well as the preceding long 3' UTR of the M gene have been reexamined in different cell lines, and it has been concluded that the former decreases F protein production, thus inhibiting MV replication and reducing cytopathogenicity, whereas the latter increases M production and promotes virus replication (Takeda et al. 2005). One can speculate that the F protein in predecessors of MV might have been much longer; when a shorter functional version of F protein arose, the remaining untranslated region, rather than being eliminated as would be dictated by the maximal compaction of genetic material in RNA viruses, became the target of biased G/I hypermutation (Cattaneo et al. 1988; Bass et al. 1989), resulting in replacement of U by C residues in the genome. (This type of massive mutational alteration has been observed frequently in defective M genes of MV triggering subacute sclerosing panencephalitis SSPE; Cattaneo et al. 1988, 1989b; Cattaneo and Billeter 1992.) Maintenance of superfluous sequences in MV might only be a somewhat realistic hypothesis in view of the virtual absence of recombination by copy choice and by the obligatory adherence to the rule of six, hallmarks of all representatives of the subfamily *Paramyxovirinae*. (For a discussion of this rule, see the following chapter.)

Another site-directed mutation was the inactivation of the C reading frame in the P gene (Radecke and Billeter 1996). This mutant propagated normally in Vero cells, most likely due to the absence of interferon type I response in these cells. This view is supported by later studies with Sendai virus, showing that ablation of C-type proteins leads to heavily compromised viruses, since Sendai C-type proteins interfere with interferon responses. In addition, these proteins are involved in transcription/replication (Delenda et al. 1998; Garcin et al. 1999, 2000, 2001, 2002). The MV C-mutant replicated poorly in human peripheral blood cells (PBMCs) (Escoffier et al. 1999) as well as in human thymus/liver implants in SCID mice (Valsamakis et al. 1998), suggesting a role of C protein also in the transcription/replication of MV.

Ablation in MV Edtag of the V protein, which arises in all naturally occurring and attenuated MV strains by co-transcriptional editing (Cattaneo et al. 1989a), or overexpression of V showed no impairment in the interferon-deficient Vero cell line (Schneider et al. 1997), nor in PBMCs (Escoffier et al. 1999), suggesting that V is not essentially required for transcription/replication. Interestingly, the V$^-$ mutant multiplied more slowly and the V overexpressing mutant more rapidly than the parent virus in human implants in SCID mice (Valsamakis et al. 1998). Furthermore, C$^-$ and V$^-$ mutants appear to be overattenuated, since they multiplied poorly in Ifnar^{ko}/CD46Ge transgenic mice, which express CD46 and are devoid of the interferon type I response (Mrkic et al. 1998). However, all these results must be interpreted with caution; in the MV Edtag strain the C-terminal portion of V protein deviating from the P protein is already defective to begin with, by replacement of important amino acids (Cys272Arg and Tyr291His) conserved in all approved MV vaccine strains (Combredet et al. 2003). Very recent studies with wild-type MV and derivatives lacking V and or C in macaques indicated that the defective MVs replicated less extensively, because they were more restricted by the interferon and inflammatory responses, whereas antibody and MV-specific T cell responses were equal for wild-type and variant viruses (Devaux et al. 2008).

The function of the matrix protein M was addressed by either eliminating most of the M reading frame (Naim et al. 2000) or by replacing the M frame in MV

Edtag by a hypermutated M reading frame recovered from the brain of a deceased SSPE patient (Patterson et al. 2001). These mutant viruses were able to replicate, albeit very slowly, in cultures of polarized cell lines derived from lung epithelia, as well as in the brain of transgenic mice expressing CD46, and in primary brain cell cultures. The experiments also showed that migration and co-localization of N and the envelope proteins F and H to particular cell compartments depend on M (Naim et al. 2000), directly demonstrating the interaction of M with nucleocapsids on one hand and with the cytoplasmic tails of the envelope proteins on the other.

Also, replacements in the MV envelope proteins were studied by directed mutations of MV strains. Replacement of single amino acids in the H protein of wild-type MV able to use only SLAM for cell entry also gained the ability to utilize CD46, the generic second receptor recognized only by attenuated vaccine strains (Erlenhofer et al. 2002). For the F protein, it has been shown that the single replacement of Leu at position 278 in the AIK-C vaccine strain by Phe, which is typical for the Schwarz/Moraten strain, resulted in enhanced syncytium formation, a distinguishing feature between the two vaccine strains (Nakayama et al. 2001).

Measles Viruses with Inserted Foreign Coding Sequences

Technical Aspects of Foreign Gene Expression

Additional transcription units (ATUs) can be inserted rather easily into the MV antigenomic sequence by placing short sequence stretches embracing an intergenic region into the 3' nontranslated region of a resident MV gene. Addition of a cassette of unique restriction sites not present in plasmids bearing antigenomic MV sequences then allows insertion of foreign reading frames (see Fig. 2 for an overview).

Care has to be taken that the final constructs obey the rule of six, which states that the number of nucleotides from the genomic 5' to the 3' terminus has to be an integral multimer of six, i.e., that the genome has to be precisely covered by N molecules, each contacting six nucleotides. This rule was first found to be valid for SeV copy-back minireplicons (Calain and Roux 1993) and then for constructed internal deletion/replacement MV minireplicons (Sidhu et al. 1995). Later, adherence to this rule was shown to be a strict requirement for full-length recombinant MV as well (Rager et al. 2002) and parainfluenzavirus type 2 (Skiadopoulos et al. 2003b) and is likely imperative for all representatives of the viral subfamily *Paramyxovirinae*. (Whenever an antigenomic cDNA construct deviates by mutation from this rule, viruses can only be rescued if they bear a second mutation preferably close to the first one, correcting the deviation.) Additional ATUs have been inserted first after the P and the H genes (Singh et al. 1999), then also before the N gene (Hangartner 1997; Zuniga et al. 2007), and recently also after the L gene (del Valle et al. 2007).

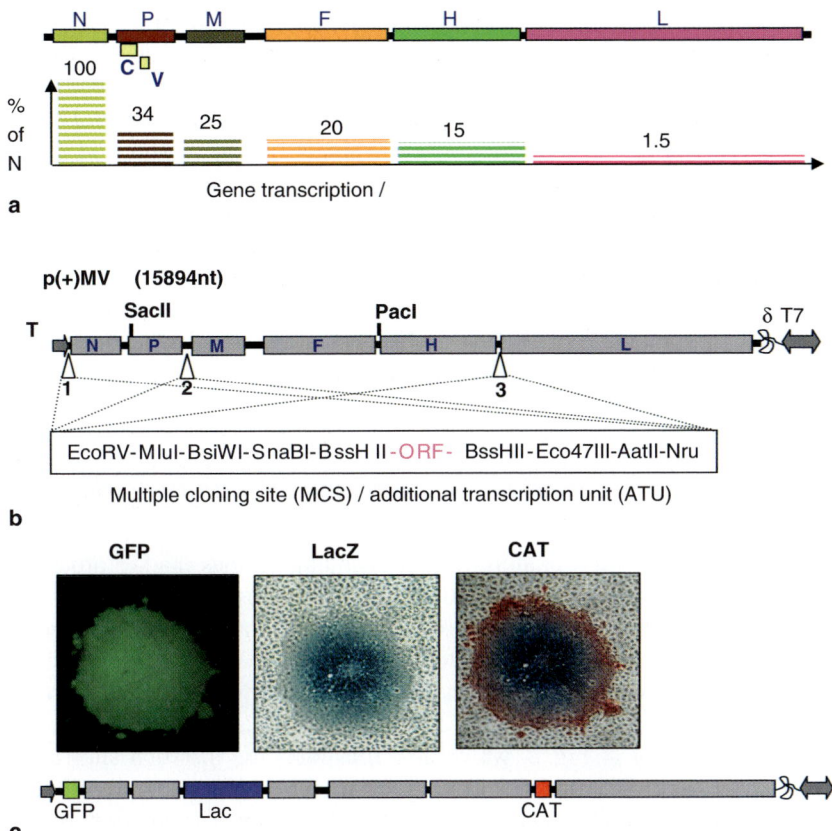

Fig. 2 Strategy for inserting transgenes into the MV backbone. **A** The MV antigenome is schematically shown, and the relative transcript levels of the six MV genes are indicated. **B** In three positions, 1 (upstream of the N gene), 2 (downstream of the P gene), and 3 (downstream of the H gene), additional transcription units (ATUs) have been inserted into antigenomic plasmids, each containing a transcription start signal, a multicloning site (MCS) embracing an open reading frame (ORF), and a termination/polyadenylation signal (Hangartner 1997; Wang et al. 2001). **C** Three different staining patterns of one syncytium is shown, induced by a triporter MV containing an eGFP ORF in position1, a beta-galactosidase (LacZ) ORF in position 2, and a CAT ORF in position 3 (Zuniga et al. 2007)

Two other approaches to express additional foreign coding sequences were also explored with the aim of ensuring that the inserted reading frames remain functionally intact, i.e., cannot be mutated without inactivating the recombinant virus. In particular, generation of viruses containing premature stop codons by the error-prone viral RNA polymerase during repeated viral passages were to be avoided. In one approach, the foreign reading frame was linked to a downstream resident MV reading frame by a stop/restart signal, by overlapping a stop codon with an adjacent initiation codon [UGAUG], as featured by influenza B viruses for the expression of the downstream NS reading frame. In the second approach, the protein self-cleavage featured by foot and mouth disease virus was adopted by linking the upstream foreign coding region with a downstream MV coding region through a stretch of nucleotides generating a very short polypeptide able to cleave almost exactly upstream of the MV protein (Ryan and Drew 1994). By this design, the transgenic protein should bear a C-terminal elongation of 17 amino acids, whereas the MV protein should bear a single foreign amino acid at its N-terminus. In both approaches, the inserts were placed upstream of the L gene, which has to be expressed only at low levels. The first method did not lead to rescuable viruses, whereas the second approach did work and might actually be employed in particular cases (L. Martin, unpublished results). However, since in a large variety of recombinant MVs mutational formation of premature stop codons was not found to actually be a problem, this approach has not been further explored in practice, at least for the time being.

Insertion of Genes Encoding Markers and Immunomodulators

It was convenient to insert reading frames encoding various marker proteins for monitoring the location of MV replication in cells (various primary cell types from infected model animals and cell lines) and in the organs of infected animals. Furthermore, the functionality of inserted ATUs can be readily ascertained, and the reading frames encoding markers can then conveniently be replaced by coding sequences of medical relevance. The most frequently used markers, apart from luciferase, are shown in Fig. 2, which also illustrates the insertion sites and the levels of expression. The most frequently used marker is green fluorescent protein (GFP) and its various derivatives (Spielhofer 1995; Duprex et al. 1999a, 1999b; Naim et al. 2000). In numerous publications not mentioned here, MV recombinants expressing markers have been used for a large variety of purposes.

Since GFP readily diffuses away from the site of its production, several attempts have been made to incorporate GFP into viral proteins, to directly visualize their location and viral assembly. Insertion into strongly expressed MV proteins has not yielded reasonably viable viruses, but rather surprisingly, one insertion into the large polymerase L was successful (Duprex et al. 2002).

Expression of cytokines, mainly to act as adjuvants to enhance the immunogenicity of vaccines, have been used in various systems. With MV, so far only IL-12 has been employed, in an attempt to completely eliminate the immunosuppression, which is the hallmark of wild-type MV infection and also characterizes vaccine strains to some extent. Functional IL-12 was expressed at high levels from inserts containing the coding regions for the two components p35 and p40 separated by an internal ribosome entry site (Singh and Billeter 1999). However, in macaques this recombinant virus failed to eliminate immunosuppression, although both humoral and cellular immune responses were altered (Hoffman et al. 2003).

Insertion of Genes from Other Pathogens: Candidate Multivalent Vaccines

Reading frames encoding proteins from other pathogens have usually been inserted into the MV context by replacement of the marker genes described above. MVs with either single or multiple insertions have been constructed, as shown only for a few examples in Fig. 3 (Zuniga et al. 2007). Most of the recombinants show replication properties in cell lines very similar to the parent empty MVs; usually the speed of propagation is only slightly delayed and the same end titers are reached. For the purposes of vaccine production, the slight delays are not counterproductive, but in the vaccinated host, a delay in propagation might result in a considerably reduced immunogenicity of the vaccine, which, however, could be counterbalanced

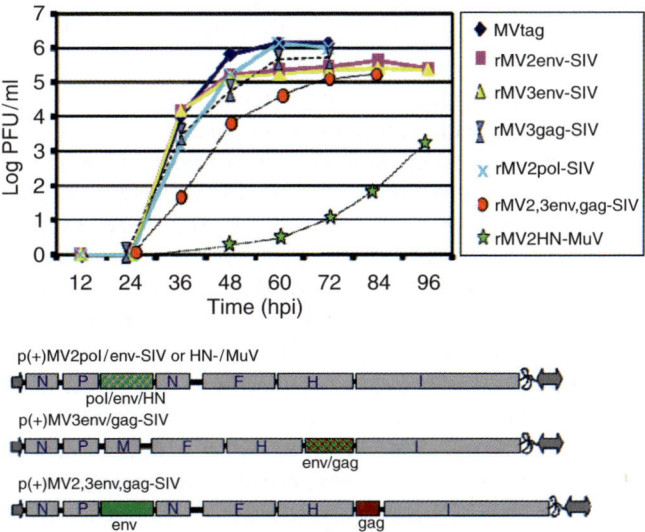

Fig. 3 Growth kinetics of selected recombinant MVs with one or two ATUs containing ORFs derived from other pathogens. Recombinant MVs expressing one or two ORFs of SIV or mumps virus were grown in comparison to the parental MV under the same physiological conditions. Growth is plotted as a function of infectious viruses released (y-axis) and time (x-axis). Experimental conditions were as described previously (Wang et al. 2001; Zuniga et al. 2007). The cloning position of all antigens is illustrated in the *lower panel*. Abbreviations: *tag* Parent empty tagged MV (without ATU), *rMV2env* or *rMV2pol* MV with a single ATU containing SIV full-length (anchored) envelope ORF or polymerase in position 2, *rMV3env* or *rMV3gag* ATU of SIV env or group-specific antigen (gag: capsid proteins) ORF in position 3, *rMV2,3env, gag* MV with 2 ATUs, containing a SIV envelope ORF in position 2 and a SIV capsid ORF in position 3, *rMV2HN-MuV*MV with a single ATU containing mumps virus hemagglutinin/neuraminidase ORF in position 2

by using higher viral doses. Given that the approved MV vaccines are administered in low doses ranging between 10^3 and 10^4 plaque forming units, this would be perfectly feasible in practice. Multiple inserts do not generally lead to more slowly growing recombinants than single inserts; the position of the inserts and the length and quality of the expressed additional proteins are much more important. Inserts at position 1 upstream of the N gene are most delaying, particularly if they are long. Glycoproteins targeted to the endoplasmic reticulum (ER) are usually more problematic than proteins remaining in the cytoplasm; furthermore, membrane-anchored glycoproteins are more deleterious than excreted ones, probably because they tend to compete with MV envelope proteins. Nevertheless, only in one instance, when MV with inserts derived from mumps virus (MuV) have been tested, was the replication delay not tolerable for practical application. This was most likely due to competition between MV and MuV glycoproteins in the assembly of viral particles (Wang et al. 2001). As evident from Fig. 3, the initial slow propagation of the recombinant expressing MuV H later resumed a speed similar to empty MV; however, the more rapidly growing virus turned out to have lost MuV H expression.

Obviously, the quality of the MV backbone is of paramount importance for the viability and thus the immunogenicity of the recombinants in vivo. As mentioned above, the MV Edtag is mutationally altered and thus exposed to the interferon type I response. (Other mutations than those in V as mentioned above, mainly in the P/C/V and the L gene, might also cripple this virus.) When tested in Ifnar^{ko}/CD46Ge mice (Mrkic et al. 1998), where the IFN-α/β receptor is inactivated, the recombinants based on this laboratory strain replicate efficiently and are reasonably immunogenic. However, in macaques, even high doses of this strain are poorly immunogenic, in contrast to the licensed Schwarz vaccine strain, which elicits strong humoral and cell-mediated immune responses (Combredet et al. 2003). It should be kept in mind that prior to 2004 in none of the published vaccination studies cited below were MV recombinants based on licensed vaccine strains used. Today, two groups (see Lorin et al. 2004; del Valle et al. 2007) use recombinants based on the Schwarz/Moraten strain obtained from different sources. (It must be recalled that the sequences of Schwarz and Moraten strains are identical, despite the reported different attenuation regimen; Parks et al. 2001a, 2001b.) Conversely, vectors based on licensed Edmonston Zagreb (EZ) vaccine strains have been developed and are being employed in several studies (H.Y. Naim et al., unpublished results).

Below, some of the pathogens addressed by recombinant MVs as described in published reports are addressed individually. In addition, a variety of projects are underway, addressing among others malaria, human papilloma virus (HPV), and avian influenza virus.

Hepatitis B Virus

Hepatitis B virus (HBV) was the first pathogen against which candidate MV-based vaccines were developed (Singh et al. 1999). The recombinant expressing the small surface antigen (HBsAg) proved to be genetically stable for at least ten serial viral transfers, a criterion subsequently checked routinely for all recombinant candidate vaccines, and induced strong humoral immune responses in Ifnar^{ko}/CD46Ge mice. A recombinant with inserts encoding both the surface and the core antigen from different cassettes in a single MV vector was also shown to express both proteins. However, this project was discontinued since development of a new HBV vaccine was deemed not to be commercially promising given that efficient vaccines against HBV already exist and that attenuated MV vaccine is not efficient in infants below the age of 6 months, whereas in developing countries chronically HBV-infected mothers usually transfer the virus to their newborns. Nevertheless, in a different laboratory, recently MV-vectored HBV vaccine candidates expressing the HBsAg have again been constructed (del Valle et al. 2007). The same insert as used previously was placed in a MV Schwarz-derived vector after the P, H, and L genes. In Ifnar^{ko}/CD46 mice, the recombinants induced antibodies; as expected, the vectors with the P and L cassettes elicited the highest and the lowest immunogenicity, respectively. In macaques, only two of the four animals inoculated with the P cassette vector generated a humoral anti-S response. It was concluded that for optimal immune responses against the insert antigens, it is important to choose the strongest expression compatible with reasonably high replication speed of the MV recombinant, or that repeat vaccinations with the same recombinant, or prime-boost regimens using either protein with adjuvant or other viral vectors should be adopted. It is noteworthy that the recombinants protected the macaques as efficiently against challenge by wild-type MV as the commercial vaccine.

It might be speculated that optimal MV-vectored HBV vaccine could even be used therapeutically for the very large number of chronically HBV-infected patients present mainly in Southeast Asia.

SIV/HIV

The target of most studies using MV vectors has been AIDS, which is arguably the most important infectious disease against which no vaccine exists as yet. Due to the very favorable record of attenuated MVs in terms of both efficacy and safety, as discussed elsewhere in this book and also in the next principal chapter, MV-vectored vaccines are among the strongest candidates for final success against AIDS. The first MV-based attempts against AIDS involved SIV antigens (Wang et al. 2001). In two laboratories, a variety of different reading frames encoding Gag, Pol, Env, and Nef derived from SIV and HIV have since been inserted, either isolated directly or manipulated in various ways, e.g., as resynthesized consensus sequences or with deletions and/or fusions. In view of the high variability of HIV, particularly env has been manipulated in different ways, to get rid of highly variable regions and thus to expose mainly conserved epitopes. In an interesting early study, anchored env appeared to be preferable to excreted forms to induce neutralizing antibodies against several heterologous HIV primary isolates and to elicit cell-mediated immune responses (Lorin et al. 2004). It is noteworthy that, as observed for HBV vaccine, MV preimmunization did not prevent these anti-HIV immune responses, as revealed in Fig. 4. In several recent reviews, the potential of MV vectored vaccines in general, and in particular against AIDS, is discussed (Tangy and Naim 2005; Zuniga et al. 2007; Brandler and Tangy 2007).

Flaviviruses: West Nile Virus, Dengue Virus

Among the human diseases caused by members of the family *Flaviviridae*, yellow fever, Japanese encephalitis, Dengue fever, and West Nile (WN) fever are the most important. Efficient vaccines exist against the first two, but morbidity and lethality caused by these diseases is still high in Western Africa/South America and in the Asia-Pacific region, respectively. In contrast, no vaccines are available against the latter two. Thus, as recently reviewed, WN and Dengue fever have been addressed using the MV-Schwarz strain (MVSchw) vector (Brandler and Tangy 2007).

The exodomains of all Flaviviruses have similar structures and are highly immunogenic; thus, rather than expressing the anchored glycoprotein as in the case of candidate AIDS vaccines, only the regions encoding the exodomains of Flaviviruses have been inserted into MV.

First, Western Nile virus (WNV) was addressed, which can efficiently infect mice by peripheral inoculation, causing a lethal neuroinvasive disease similar to encephalitis in humans. The secreted form of the glycoprotein E (sE) was expressed, and low doses of MVSchw-sEWNV protected mice against lethal challenges by WNV at least for a period ranging from 8 days to 6 months after vaccination (Despres et al. 2005). In currently unpublished experiments, the efficacy of the recombinant was also demonstrated in a primate model: squirrel monkeys (*Saimiri sciureus*) were completely protected against WNV by a single-dose vaccination.

Dengue virus (DV) poses a greater problem, as four serotypes exist. The exposed portion of gpE folds into three globular ectodomains ED I, II, and III; the outermost immunoglobulin-like ED III, which can fold independently through a single disulfide bond, determines the serotype specificity and contains the major neutralizing epitopes. Thus, as announced by the Tangy group (Brandler et al. 2007), a construct encoding all four EDIII domains of DV 1, 2, 3, and 4 has been inserted into the MVSchw vector, and immunogenicity studies are ongoing.

SARS Corona Virus

Immediately following the SARS epidemic, and because of the fear that SARS Corona virus (CoV) might be used for bioterrorism, both the nucleocapsid protein N and the codon-optimized spike glycoprotein S of SARS-CoV were inserted independently into position 2 between the P and M gene of MV Edtag, yielding p(+)MV-SARS-N and p(+)-SARS-S. In MV-susceptible Ifnar^{ko}/CD46Ge mice, the recombinants induced high antibody titers against N and S. The antibodies generated against the S-expressing MV strongly neutralized both SARS-CoV and MV. N-expressing MV elicited a cellular immune response measured by IFN-γ ELISPOT. Also in this case, preimmunization with empty MV did not impair the immune responses against SARS CoV, and a 1:1 mixture of the two recombinants induced responses of a similar magnitude to that of the single vaccines (Liniger et al. 2008).

Creation of Targeted MVs with Oncolytic Properties

The projects to develop recombinant MVs against cancer are extensively discussed elsewhere in this book (see the chapter by S. Russel et al., this Volume). Here it should only be mentioned that these oncolytic viruses are constructed and rescued basically as the recombinants discussed above. Oncolytic properties of MV have been recognized by the early finding that MV infection resolved Burkitt's lymphoma (Bluming and Ziegler 1971). To enhance these properties, numerous alterations have been introduced into the MV derived from vaccines, with the intent of retargeting the recombinants specifically to cancer cells and ablating interaction with the known MV receptors SLAM and CD46. Thus, primarily the H reading frame has been manipulated to create slightly shortened H proteins carrying at their C-termini ligands fitting onto different receptors expressed abundantly by diverse types of cancer cells. The first promising recombinants have been created by the authors of a recent review (von Messling and Cattaneo 2004; for later original publications see Springfeld et al. 2006; Ungerechts et al. 2007a, 2007b).

Development of Segmented MVs Carrying Foreign Genes

One of the most interesting developments concerning manipulated MVs is the recent creation of MVs containing bi- or even tri-partite genomes, denominated 2 seg-MV and 3 seg-MV, respectively (Takeda et al. 2006). The DNA constructs underlying these segmented viruses were designed to bear precisely the 3' and 5' terminal untranslated regions present in the parent MV. In addition to the MV genes, the 3 seg-MV could accommodate up to six additional transcription units, and efficient expression of five of the encoded proteins was demonstrated. Importantly, expression of a long insert encoding beta-galactosidase placed upstream of the N gene in a segment comprising only the MV N and P genes impaired the replication of the recombinant much less than beta-galactosidase residing in the same position of a monopartite MV. This suggests that long, multiple inserts could be expressed more efficiently when segmented MVs are used as vectors. The surprisingly efficient propagation can be explained by the fact that the MV particles grown in cell lines are largely polyploid (Rager et al. 2002), containing far more than one genome. In addition, when relatively short genomes are replicated, the probability of the replicase to quit the template prematurely is lower than on long templates.

Until such segmented MVs can be used for evaluation in practice, much additional experimentation is needed. First, an attenuated licensed MV vaccine strain has to be used rather that the wild-type isolate from which the constructs were derived. Second, a rescue method has to be used employing exclusively transfected plasmids rather than vaccinia-T7 helper virus. Third, for the construction of tripartite virus, each segment should contain one of the genes required for replication, N, P, or L. (Ideally, to fully exploit the potential of segmented MV vectors, one segment should bear only one short transgene upstream of the measles genes encoding M, F, H, and L, whereas two segments encoding N and P, respectively, could bear multiple and relatively long inserts up- and downstream of the MV genes.) Fourth, propagation and immunogenicity in animal models have to be checked in comparison to monopartite MV vectors. These provisos by no means reduce the importance of this pioneering work.

Considerations for Practical Applications of Recombinant Multivalent MV Vaccines

Stability of Recombinant MVs

RNA polymerases are known to be much more error-prone than DNA polymerases due to the lack of proofreading; thus, the error rate in RNA viruses is on average approximately three orders of magnitude higher than in DNA viruses. The so-called quasispecies concept for RNA viruses was explored in an early study of bacteriophage Qbeta (Domingo et al. 1978). Therefore, it came as a great surprise to find that in the MV recombinants the transgenes are rather stably expressed, although they are nonessential and in some cases deleterious for virus propagation and expected to be even more rapidly mutated than the MV genes conserved by positive selection. The expression of the first transgene studied in MV encoding chloramphenicol acetyl transferase (CAT), inserted after the P and H gene, was completely maintained in all analyzed progeny clones derived from single original rescued clones after ten serial virus passages, which represent an amplification factor of 10^{33} (Spielhofer 1995). However, in that experiment, 35% and 100% of the clones derived from second rescued clones serially transferred in parallel had lost CAT expression. It seems likely that the errors already occurred in these second original rescued clones during transcription by the particularly error-prone T7 RNA polymerase. In almost all later rescued recombinants, the transgene expression was maintained. Only the expression of MuV-derived transgenes encoding MuV F and HN was lost, resulting in viruses propagating as fast as empty MV, whereas the freshly rescued recombinants were slow, as discussed above and shown in Fig. 3.

In none of these few cases where transgene expression was lost could any deletion be detected as monitored by PCR, suggesting that recombination by copy-choice must be an extremely rare event, in contrast to plus-strand RNA viruses, which readily delete inserted foreign sequences. Copy-choice events must occur sometimes in MV, as the generation of defective interfering particles (DIs) cannot arise in any other fashion. However, most of the DIs are the copy-back type, where the polymerase, after leaving the original template together with the unfinished transcript, attaches to this transcript rather than to the original template. Virtual absence of deletion events must be due to the extremely tight RNP structure, which renders reattachment of the replicase to its template in another position almost impossible.

Misincorporation of single nucleotides certainly does occur in MV to an appreciable extent, as documented by nucleotide changes in characteristic positions during the attenuation process to obtain vaccine strains (Parks et al. 2001a, 2001b) and by loss of virulence upon passage of wild-type virus isolates in nonlymphoid cells (Kobune et al. 1990; Tatsuo et al. 2000). However, in these cases, the mutational changes resulted in a propagation advantage in the cells used for attenuation and were thus selected for against the original virus. In contrast, loss of expression of a protein which does not interfere with viral propagation is neutral. Nevertheless, it is surprising that ablation of protein expression, which could easily arise by nucleotide exchanges, leading to premature stop codons and thus interruption of the coding region, has not been documented. Single nucleotide deletions and insertions, which could also lead to truncation of the expressed protein, are virtually excluded due to the strict adherence of MV and all other *Paramyxovirinae* to the rule of six discussed above. Missense mutation in transgenes are obviously more difficult to detect than nonsense mutations, and to our knowledge for none of the inserts in MV have such events been excluded by sequencing after ten standard serial viral transfers. However, as long as the antigenic properties of the expressed protein are not severely hampered, in most cases such changes are expected to deteriorate the recombinant vaccine only very slightly.

Efficacy and Safety

The efficacy of vaccination using attenuated MVs is quite remarkable: most vaccinees seroconvert and are then protected for life, at least against clinically manifest MV infection. At present, it is only a hope that MV-expressed single antigens of other pathogens will elicit a comparable strong and durable protection. Nevertheless, it is encouraging that in the first instance where protection of an animal model close to humans has been tested, in the case of MV expressing the exodomain of WNV gpE, a complete protection up to 6 months has been documented in monkeys, as mentioned above (Brandler et al. 2007). Clinical trials with any recombinant vaccine candidate based on MV are only in the planning stage; thus, practical success is still to be awaited for some years to come in view of the severe and costly hurdles imposed by regulatory agencies.

Nevertheless, the potential of MV recombinants as efficient multivalent vaccines has started to be appreciated by increasing numbers of vaccination experts. Even if in particularly difficult cases such as malaria, elicited by several different and variable strains of *Plasmodium*, complete protection cannot be expected, MV recombinants might elicit immune responses comparable to that triggered by the first *Plasmodium* infection that every infant in sub-Saharan Africa contracts. This would already be an important success, since the first infection is responsible for more than 80% of deaths caused by malaria, whereas further *Plasmodium* infections involve much less morbidity. The impact of MV recombinants on AIDS might be considered less optimistically, since a partially protecting vaccine would most likely result in increased negligence of other protective measures.

How does MV compare with other vaccination regimens based on delivery of genetic material? In comparison with plasmid DNA, the efficacy is expected to be much higher, and in terms of safety it is certainly favorable that the danger of integration of genetic material into the human genome is negligible in comparison to delivered DNA or to DNA viruses. This expectation appears valid despite reported integration of cDNA derived from lymphocytic choriomeningitis virus (LCMV) into mouse genomes (Klenerman et al. 1997). An additional advantage is the relatively small size of MV in comparison with DNA viruses used as delivery vehicles; only a few proteins of the vector are expressed in addition to those expressed from the artificial inserts, thus minimizing superfluous immune responses. In fact, the immune responses against the vector must also be considered beneficial, at least as long as MV is not eradicated, a goal of the WHO that will take many years to achieve and might never be reached. (Here it should be mentioned that in the opinion of some medical experts of MV vaccination in Africa, attenuated MV vaccination is generally beneficial for the development of the immune system in infants and therefore should be maintained even if measles is eradicated one day.) In comparison with RNA virus vectors, the high genetic stability of MV is noteworthy, which is shared with the representatives of the viral subfamily *Paramyxovirinae*, all representatives of which adhere to the rule of six, whereas the other *Mononegavirales* are somewhat less stable. In addition, the high possible payload of inserts, unrestricted by viral capsid structure, should be noted.

What are the consequences of MV vaccines being replication-competent? Replication-deficient vectors such as most DNA viruses in use and RNA virus vectors based on alpha-viruses are broadly considered to be generally safer. However, replication-deficiency entails two major drawbacks. On one hand, doses higher by many orders of magnitude in comparison with replicating vehicles have to be delivered. Even more importantly, the range of action is very limited, as only one-step infection of cells bearing suitable virus receptors at the place of delivery is possible. This contrasts significantly with attenuated MV vaccines, which owe their high efficacy to systemic spread and preferential infection of professional antigen-presenting cells and lymphoid tissues. In the case of MV replication, competence appears only as an advantage in terms of safety and efficacy, since MV vaccines have an extremely good safety record (MV vaccination is recommended even for immunocompromised patients, e.g., HIV), and severe sequelae of wild-type MV infection such as SSPE have been stopped by vaccination. Importantly, cell targeting of MV recombinants is not altered by transgenes (no envelope proteins, except MV-related glycoproteins as those of the MuV envelope have been shown to be incorporated into MV particles; Wang et al. 2001).

Quite generally, it is noteworthy that killed vaccines or subunit vaccines are not necessarily safer than replication-competent vaccines, as shown for killed MV vaccine, which proved not only to lack efficacy, but even exacerbated the effects of subsequent wild-type MV infection. Thus, there is no general increasing gradient of safety when one compares replication-competent versus replication-restricted versus dead vaccines. Any component delivered to humans must be planned, developed, and evaluated with equal care, without a priori bias based on theoretical general and illusive classifications of danger to human health.

Preimmunity: Prime-Boost Regimens

Preimmunity is a concern for the use of recombinant MV vaccines given that (a) almost all adolescent and adult individuals are immune against measles due to prior wild-type MV infection or MV vaccination and (b) maternal antibodies against measles in young infants below 6–9 months of age prevent immunization by attenuated MV vaccines. Therefore, vaccination with recombinant MVs was initially considered exclusively for young infants at an age when MV vaccination is recommended, thus replacing the standard vaccine with a bi- or multivalent vaccine protecting not only against measles, but also against other important diseases such as malaria. Such a regimen is certainly still a preferred option, and care has been taken to ascertain that in animal models various transgenes in MV do not hamper the immunogenicity against measles (Brandler et al. 2007; Brandler and Tangy 2007; del Valle et al. 2007). However, measles immunity can be boosted in children who are preimmune after a first vaccination, particularly when the boost is given as an aerosol (Bennett et al. 2002). Thus, it was probably wrong to assume that the immune reactions against measles in young infants, where the immune system is still immature, is necessarily similar in older children and adults. Since, for example, MV-based AIDS vaccines should be given much later, before sexual activity begins, it was important to ascertain, for the time being in animal models, that preimmunity from vaccination with standard MV vaccine does not prevent a boost with recombinant MV. In fact, boosts resulted in enhancement of immune responses against MV and induction of immune responses against the transgenes (del Valle et al. 2007; Brandler and Tangy 2007; see Fig. 4).

Nevertheless, efficiency of boosting with MV recombinants will probably wane with increasing numbers of boosts. Thus, prime-boost regimens using either subunit

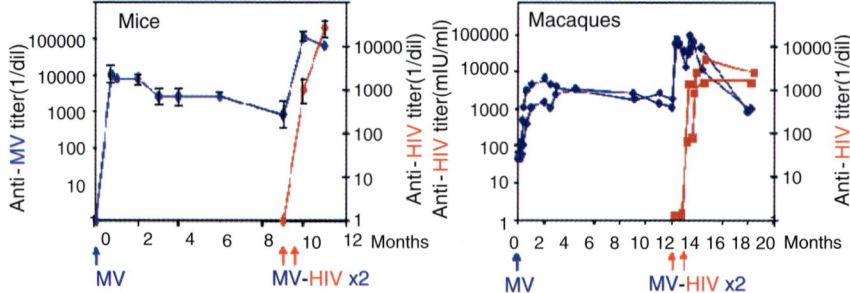

Fig. 4 Induction of immune responses against antigen ORFs in recombinant MV in presence of anti-MV antibodies. Antibody titers against MV (*blue*) and against HIV (*red*) in mice and macaques. *Left panel* A group of six transgenic CD46^{ko}/IfnarGe mice (Mrkic et al. 1998) were preimmunized with 10^5 TCID$_{50}$ of MV Edtag and 9 months later twice with 5×10^6 TCID$_{50}$ of recombinant MV containing the HIV envelope glycoprotein gp140$_{HIV89.6}$ in position 2. *Right panel* Two cynomolgus macaques were preimmunized with a standard human vaccination dose (10^4 TCID$_{50}$) of commercial Rouvax MV, and immunized at months 12 and 13 with the same recombinant utilized for the mice, using 5×10^6 TCID$_{50}$ (Lorin et al. 2004; Tangy and Naim 2005). (Modified with permission from Brandler and Tangy 2007)

vaccines with adjuvants or delivery with other viral vectors may be advantageous in practice. Adenovirus vectors should induce immune responses in newborns already and might close the gap until MV vaccines can be used. A particularly attractive regimen might entail vectors based on attenuated MuV, which, similar to MV, is being used widely for human vaccination. Furthermore, other vectors based on *Mononegavirales* and among those preferentially belonging to the *Paramyxovirinae* should also be considered. Representatives of all genera belonging to this subfamily and thus far unclassified viruses might be explored, as no cross-reactivity between these genera exists. However, attenuated MV and MuV are the only viruses for which extensive experience in humans exists, and are therefore preferred options for vector use, particularly since they are able, at least in principle, to induce life-long immunity.

Impediments Imposed by Regulatory and Commercial Issues

It is sad that the progress toward new vaccines and the improvement of existing vaccines is blocked or severely slowed down by issues other than those based on scientific and technical considerations. This is true in general: just one example is the failure to improve the Sabin vaccine against polio by planned mutations, which would impede reversion to wild-type viruses that repeatedly led to small epidemics. Such impediments are highly regrettable in the case of MV and similar viruses for vector use, which appear promising in the battle against important diseases. Vaccines developed on the basis of such vectors may particularly enhance public health in developing countries, considering the reasons outlined above and the low costs of production per vaccination dose.

Regulatory Hurdles

The great importance for human health and the high level of responsibility expected from regulatory agencies such as the FDA in the United States is fully recognized. Nevertheless, it is difficult to understand why, in cases such as the polio vaccine mentioned in the preceding section, phase III clinical trials costing hundreds of millions of dollars would be imposed even for applying minor changes that would render the vaccine much safer; one cannot blame manufacturers for refraining from improving their product in view of the enormous costs entailed. Similarly, imposing such extensive and costly clinical trials might not be warranted for recombinants based on the widely used MV vaccines, provided that it can be rigorously demonstrated that any of the additionally expressed antigen is not toxic, the cell targeting of the recombinant virus is not altered by inserted transgenes, the immune response generated against measles is the same as in the traditional MV vaccines, and the manufacturing process is practically the same as that used for the traditional vaccine.

It might also be asked whether society should not generally accept the prevailing principle of ethical behavior, i.e., maximizing the average benefits for the largest possible number of individuals. Risks accompany every human invention, and in many cases, such as traffic and sport activities, risks are tacitly accepted. Why is this not the case with vaccines? Rare adverse effects of medically active compounds, as routinely explained in leaflets accompanying drugs, are usually tolerated. Admittedly, vaccines must be considered differently, as they are administered to healthy individuals, particularly to children. Nevertheless, why should even highly adverse effects of vaccines not be acceptable, as long as such effects remain very rare, when severe morbidity and death can be prevented for large populations? Fact-based, nonemotional dialogue between scientists and society is clearly necessary. It should be possible to find acceptable legal and procedural solutions to this and similar problems even though at present in general the rights of the individual prevail over the rights of the community.

Patent Conflicts and Disruption of Scientific Collaborations

Unfortunately, patents are unavoidable to actually produce, based on scientific endeavors either in academic or corporate research laboratories, medically valuable commercial products. Competition between companies and scientists is an important driving force and may be considered more beneficial than detrimental in the long run. However, for the development of *Mononegavirales* as vaccine vectors, many researchers in different laboratories have invented tools that are indispensable to the success in establishing reverse genetics and foreign gene expression. In this instance, competition and financial avidity not only among companies, but also among scientists, has severely hampered progress. Patent offices must necessarily adhere to legal considerations, but research institutions should be amenable to compromises based on the recognition of scientific merit. Patent conflicts are costly and demotivating in general. They become intolerable when, for example, scientific collaborations that have arisen over many years and are subsidized by public grant money are unilaterally discontinued by decree of an institutional director, forcing his researchers, for the financial benefit of his institution, to discontinue the flow of materials and information.

Other Recombinant *Mononegavirales* as Vaccines

As discussed in various sections above, a variety of *Mononegavirales* have been rescued from DNA. The first reported rescues of a number of these viruses are addressed in an excellent general review on *Paramyxoviridae* and their interactions with the host (Lamb and Parks 2007). The list includes RABV (Schnell et al. 1994), VSV (Lawson et al. 1995; Whelan et al. 1995), MV (Radecke et al. 1995), human respiratory virus hRSV (Collins et al. 1995), SeV (Garcin et al. 1995; Kato et al. 1996), rinderpest virus (Baron and Barrett 1997), hPIV3 (Durbin et al. 1997; Hoffman and Banerjee 1997), SV5/PIV5 (He et al. 1997), NDV (Peeters et al. 1999), and bRSV (Buchholz et al. 1999).

Some examples of foreign gene insertions in *Paramyxoviridae* are reported in the review on Parainfluenza viruses, which comprise the genera *Respirovirus*, *Rubulavirus*, and *Avulavirus* (Karron and Collins 2007). The bPIV3 backbone bearing the hPIV3 F and HN genes rather than their bovine counterparts (Schmidt et al. 2000), and hPIV in which only a few internal protein genes such as N and P are replaced by their bovine counterparts (Bailly et al. 2000; Skiadopoulos et al. 2003a) are candidate vaccines against human parainfluenza. Additional foreign genes derived from other viruses such as hRSV and human metapneumovirus (hMPV) have also been inserted into hPIV3 (Schmidt et al. 2002; Skiadopoulos et al. 2002), as well as in hPIV1, hPIV2, and SeV. Vectors based on a representative of the genus *Avulavirus*, Newcastle disease virus (NDV, now officially called AMPV1) appear promising as vaccines and oncolytic agents for veterinarian and human use (Ge et al. 2007; Lorence et al. 2007; DiNapoli et al. 2007; Estevez et al. 2007; Vigil et al. 2007; Janke et al. 2007). A drawback for respiroviruses as vaccination vectors is that even a full infection with these nonattenuated viruses does not protect from reinfection by the same agent. Thus, protection by attenuated respirovirus vectors cannot be expected to be long-lasting.

Of the two main genera *Lyssavirus* and *Vesiculovirus* of the recently reviewed *Rhabdoviridae* (Lyles and Rupprecht 2007), rabies virus (RABV) and the two main subtypes of VSV (Indiana and New Jersey) have been proposed both as vaccination vectors, mainly against AIDS, and as therapeutic agents (Roberts et al. 1999; Roberts and Rose 1999; Rose et al. 2001; Schnell et al. 2000; Palin et al. 2007; Simon et al. 2007; Clarke et al. 2006; Bergman et al. 2007; Diaz et al. 2007; Brandsma et al. 2007; Schwartz et al. 2007; Tani et al. 2007; Cooper et al. 2008). RABV, and particularly VSV and derived recombinants containing inserts, have the advantage of replicating to extremely high titers. Since the natural hosts of VSV are cattle, horses, swine, mosquitoes, and sandflies, human use is somewhat problematic, particularly since VSV shows some neurovirulence in primates (Johnson et al. 2007). However, VSV can be greatly attenuated by shuffling its genes (Wertz et al. 1998).

Conclusions

In comparison with other *Mononegavirales*, in particular VSV and NDV/AMPV1, MV vectors have lagged behind, as is evident from the recently published reports cited in the preceding section. In part, this may stem from the relatively low titers obtained for MVs, to preexisting immunity against MV in the adolescent and adult human population, as discussed above, and suboptimal and/or expensive animal models. Clinical studies involving MV vectors for vaccine applications are only in the planning phase and are presently ongoing only with oncolytic MV recombinants requiring less strict previous testing. Thus, the promise of recombinant MV vaccines obtained in studies with animal models has yet to be corroborated in humans, the unique MV host. Nevertheless, for a variety of reasons discussed in this chapter, it can be predicted that the search for medical applications with recombinant MVs will rapidly gain momentum, as suggested by recent reviews comparing different viral vectors for vaccination purposes (Li et al. 2007; Brave et al. 2007).

In comparison with other vectors based on RNA viruses, MV and generally *Mononegavirales* appear clearly superior. Will the hope presently attached to these vectors gradually wane, as has occurred for many DNA virus applications and delivered plasmid DNA? Comparative studies are clearly needed, as are studies comparing replication restricted to replication-competent vectors, which are thought to be superior at least in case of MV and MuV.

Acknowledgements Support for the experimental work carried out in the authors' laboratories was provided by grants from the Swiss National Science Foundation (SNSF), the Swiss Commission for Technical Innovation (KTI), the Italian Government, the European Commission, and the NIH-NIAID. We gratefully acknowledge privileged communication of results prior to publication by Roberto Cattaneo, Mayo Clinic, Rochester, MN USA; Frédéric Tangy, Institut Pasteur, Paris, France; Yusuke Yanagi, Kyushu University, Fukuoka, Japan, and Urs Schneider, University of Freiburg, Germany. We are also indebted to Urs Schneider for providing plasmids containing the chicken actin (CA) promoter. Conflicting interests: MAB and SAU are principal inventors of patent requests filed for property of Berna Biotech Inc., Berne, Switzerland (a Crucell Company), and Weyeth-Lederle Vaccines, Pearl River, New York 10965, USA, respectively. They do not receive any personal financial benefits from granted patents.

References

Andino R, Silvera D, Suggett SD, Achacoso PL, Miller CJ, Baltimore D, Feinberg MB (1994) Engineering poliovirus as a vaccine vector for the expression of diverse antigens. Science 265:1448–1451

Arroyo J, Miller C, Catalan J, Myers GA, Ratterree MS, Trent DW, Monath TP (2004) ChimeriVax-West Nile virus live-attenuated vaccine: preclinical evaluation of safety, immunogenicity, and efficacy. J Virol 78:12497–12507

Atkins GJ, Sheahan BJ, Liljestrom P (1996) Manipulation of the Semliki Forest virus genome and its potential for vaccine construction. Mol Biotechnol 5:33–38

Bailly JE, McAuliffe JM, Durbin AP, Elkins WR, Collins PL, Murphy BR (2000) A recombinant human parainfluenza virus type 3 (PIV3) in which the nucleocapsid N protein has been replaced by that of bovine PIV3 is attenuated in primates. J Virol 74:3188–3195

Baron MD, Barrett T (1997) Rescue of rinderpest virus from cloned cDNA. J Virol 71:1265–1271

Bass BL, Weintraub H, Cattaneo R, Billeter MA (1989) Biased hypermutation of viral RNA genomes could be due to unwinding/modification of double-stranded RNA. Cell 56:331

Bennett JV, Fernandez de Castro J, Valdespino-Gomez JL, Garcia-Garcia Mde L, Islas-Romero R, Echaniz-Aviles G, Jimenez-Corona A, Sepulveda-Amor J (2002) Aerosolized measles and measles-rubella vaccines induce better measles antibody booster responses than injected vaccines: randomized trials in Mexican schoolchildren. Bull World Health Organ 80:806–812

Bergman I, Griffin JA, Gao Y, Whitaker-Dowling P (2007) Treatment of implanted mammary tumors with recombinant vesicular stomatitis virus targeted to Her2/neu. Int J Cancer 121:425–430

Bluming AZ, Ziegler JL (1971) Regression of Burkitt's lymphoma in association with measles infection. Lancet 2:105–106

Brandler S, Tangy F (2007) Recombinant vector derived from live attenuated measles virus: potential for flavivirus vaccines. Comp Immunol Microbiol Infect Dis 31:271–291

Brandler S, Brown N, Ermak TH, Mitchell F, Parsons M, Zhang Z, Lang J, Monath TP, Guirakhoo F (2005) Replication of chimeric yellow fever virus-dengue serotype 1–4 virus vaccine strains in dendritic and hepatic cells. Am J Trop Med Hyg 72:74–81

Brandler S, Lucas-Hourani M, Moris A, Frenkiel MP, Combredet C, Fevrier M, Bedouelle H, Schwartz O, Despres P, Tangy F (2007) Pediatric measles vaccine expressing a Dengue antigen induces durable serotype-specific neutralizing antibodies to Dengue virus. PLoS Negl Trop Dis 1:e96

Brandsma JL, Shylankevich M, Su Y, Roberts A, Rose JK, Zelterman D, Buonocore L (2007) Vesicular stomatitis virus-based therapeutic vaccination targeted to the E1, E2, E6, and E7 proteins of cottontail rabbit papillomavirus. J Virol 81:5749–5758

Brave A, Ljungberg K, Wahren B, Liu MA (2007) Vaccine delivery methods using viral vectors. Mol Pharm 4:18–32

Buchholz UJ, Finke S, Conzelmann KK (1999) Generation of bovine respiratory syncytial virus (BRSV) from cDNA: BRSV NS2 is not essential for virus replication in tissue culture, and the human RSV leader region acts as a functional BRSV genome promoter. J Virol 73:251–259

Calain P, Roux L (1993) The rule of six, a basic feature for efficient replication of Sendai virus defective interfering RNA. J Virol 67:4822–4830

Cattaneo R, Billeter MA (1992) Mutations and A/I hypermutations in measles virus persistent infections. Curr Top Microbiol Immunol 176:63–74

Cattaneo R, Schmid A, Eschle D, Baczko K, ter Meulen V, Billeter MA (1988) Biased hypermutation and other genetic changes in defective measles viruses in human brain infections. Cell 55:255–265

Cattaneo R, Kaelin K, Baczko K, Billeter MA (1989a) Measles virus editing provides an additional cysteine-rich protein. Cell 56:759–764

Cattaneo R, Schmid A, Spielhofer P, Kaelin K, Baczko K, ter Meulen V, Pardowitz J, Flanagan S, Rima BK, Udem SA et al (1989b) Mutated and hypermutated genes of persistent measles viruses which caused lethal human brain diseases. Virology 173:415–425

Clarke DK, Sidhu MS, Johnson JE, Udem SA (2000) Rescue of mumps virus from cDNA. J Virol 74:4831–4838

Clarke DK, Cooper D, Egan MA, Hendry RM, Parks CL, Udem SA (2006) Recombinant vesicular stomatitis virus as an HIV-1 vaccine vector. Springer Semin Immunopathol 28:239–253

Collins PL, Mink MA, Stec DS (1991) Rescue of synthetic analogs of respiratory syncytial virus genomic RNA and effect of truncations and mutations on the expression of a foreign reporter gene. Proc Natl Acad Sci U S A 88:9663–9667

Collins PL, Mink MA, Hill MG 3rd, Camargo E, Grosfeld H, Stec DS (1993) Rescue of a 7502-nucleotide (49.3% of full-length) synthetic analog of respiratory syncytial virus genomic RNA. Virology 195:252–256

Collins PL, Hill MG, Camargo E, Grosfeld H, Chanock RM, Murphy BR (1995) Production of infectious human respiratory syncytial virus from cloned cDNA confirms an essential role for the transcription elongation factor from the 5′ proximal open reading frame of the M2 mRNA in gene expression and provides a capability for vaccine development. Proc Natl Acad Sci U S A 92:11563–11567

Combredet C, Labrousse V, Mollet L, Lorin C, Delebecque F, Hurtrel B, McClure H, Feinberg MB, Brahic M, Tangy F (2003) A molecularly cloned Schwarz strain of measles virus vaccine induces strong immune responses in macaques and transgenic mice. J Virol 77:11546–11554

Conzelmann KK, Schnell M (1994) Rescue of synthetic genomic RNA analogs of rabies virus by plasmid-encoded proteins. J Virol 68:713–719

Cooper D, Wright KJ, Calderon PC, Guo M, Nasar F, Johnson JE, Coleman JW, Lee M, Kotash C, Yurgelonis I, Natuk RJ, Hendry RM, Udem SA, Clarke DK (2008) Attenuation of recombinant vesicular stomatitis virus-human immunodeficiency virus type 1 vaccine vectors by gene translocations and g gene truncation reduces neurovirulence and enhances immunogenicity in mice. J Virol 82:207–219

del Valle JR, Devaux P, Hodge G, Wegner NJ, McChesney MB, Cattaneo R (2007) A vectored measles virus induces hepatitis B surface antigen antibodies while protecting macaques against measles virus challenge. J Virol 81:10597–10605

Delenda C, Taylor G, Hausmann S, Garcin D, Kolakofsky D (1998) Sendai viruses with altered P, V, and W protein expression. Virology 242:327–337

Despres P, Combredet C, Frenkiel MP, Lorin C, Brahic M, Tangy F (2005) Live measles vaccine expressing the secreted form of the West Nile virus envelope glycoprotein protects against West Nile virus encephalitis. J Infect Dis 191:207–214

Devaux P, Hodge G, McChesney MB, Cattaneo R (2008) Attenuation of V- or C-defective measles viruses: infection control by the inflammatory and interferon responses of rhesus monkeys. J Virol 82:5359–5367

Diaz RM, Galivo F, Kottke T, Wongthida P, Qiao J, Thompson J, Valdes M, Barber G, Vile RG (2007) Oncolytic immunovirotherapy for melanoma using vesicular stomatitis virus. Cancer Res 67:2840–2848

DiNapoli JM, Kotelkin A, Yang L, Elankumaran S, Murphy BR, Samal SK, Collins PL, Bukreyev A (2007) Newcastle disease virus, a host range-restricted virus, as a vaccine vector for intranasal immunization against emerging pathogens. Proc Natl Acad Sci U S A 104:9788–9793

Domingo E, Sabo D, Taniguchi T, Weissmann C (1978) Nucleotide sequence heterogeneity of an RNA phage population. Cell 13:735–744

Duprex WP, Duffy I, McQuaid S, Hamill L, Cosby SL, Billeter MA, Schneider-Schaulies J, ter Meulen V, Rima BK (1999a) The H gene of rodent brain-adapted measles virus confers neurovirulence to the Edmonston vaccine strain. J Virol 73:6916–6922

Duprex WP, McQuaid S, Hangartner L, Billeter MA, Rima BK (1999b) Observation of measles virus cell-to-cell spread in astrocytoma cells by using a green fluorescent protein-expressing recombinant virus. J Virol 73:9568–9575

Duprex WP, Collins FM, Rima BK (2002) Modulating the function of the measles virus RNA-dependent RNA polymerase by insertion of green fluorescent protein into the open reading frame. J Virol 76:7322–7328

Durbin AP, Hall SL, Siew JW, Whitehead SS, Collins PL, Murphy BR (1997) Recovery of infectious human parainfluenza virus type 3 from cDNA. Virology 235:323–332

Enami M, Luytjes W, Krystal M, Palese P (1990) Introduction of site-specific mutations into the genome of influenza virus. Proc Natl Acad Sci U S A 87:3802–3805

Erlenhofer C, Duprex WP, Rima BK, ter Meulen V, Schneider-Schaulies J (2002) Analysis of receptor (CD46, CD150) usage by measles virus. J Gen Virol 83:1431–1436

Escoffier C, Manie S, Vincent S, Muller CP, Billeter M, Gerlier D (1999) Nonstructural C protein is required for efficient measles virus replication in human peripheral blood cells. J Virol 73:1695–1698

Estevez C, King D, Seal B, Yu Q (2007) Evaluation of Newcastle disease virus chimeras expressing the Hemagglutinin-Neuraminidase protein of velogenic strains in the context of a mesogenic recombinant virus backbone. Virus Res 129:182–190

Fodor E, Devenish L, Engelhardt OG, Palese P, Brownlee GG, Garcia-Sastre A (1999) Rescue of influenza A virus from recombinant DNA. J Virol 73:9679–9682

Garcin D, Pelet T, Calain P, Roux L, Curran J, Kolakofsky D (1995) A highly recombinogenic system for the recovery of infectious Sendai paramyxovirus from cDNA: generation of a novel copy-back nondefective interfering virus. EMBO J 14:6087–6094

Garcin D, Latorre P, Kolakofsky D (1999) Sendai virus C proteins counteract the interferon-mediated induction of an antiviral state. J Virol 73:6559–6565

Garcin D, Curran J, Kolakofsky D (2000) Sendai virus C proteins must interact directly with cellular components to interfere with interferon action. J Virol 74:8823–8830

Garcin D, Curran J, Itoh M, Kolakofsky D (2001) Longer and shorter forms of Sendai virus C proteins play different roles in modulating the cellular antiviral response. J Virol 75:6800–6807

Garcin D, Marq JB, Strahle L, le Mercier P, Kolakofsky D (2002) All four Sendai Virus C proteins bind Stat1, but only the larger forms also induce its mono-ubiquitination and degradation. Virology 295:256–265

Ge J, Deng G, Wen Z, Tian G, Wang Y, Shi J, Wang X, Li Y, Hu S, Jiang Y, Yang C, Yu K, Bu Z, Chen H (2007) Newcastle disease virus-based live attenuated vaccine completely protects chickens and mice from lethal challenge of homologous and heterologous H5N1 avian influenza viruses. J Virol 81:150–158

Hangartner L (1997) Development of measles virus as a vector: expression of green fluorescent protein from different loci. Master's thesis, University of Zurich, Zurich, Switzerland

He B, Paterson RG, Ward CD, Lamb RA (1997) Recovery of infectious SV5 from cloned DNA and expression of a foreign gene. Virology 237:249–260

Hoffman MA, Banerjee AK (1997) An infectious clone of human parainfluenza virus type 3. J Virol 71:4272–4277

Hoffman SJ, Polack FP, Hauer DA, Singh M, Billeter MA, Adams RJ, Griffin DE (2003) Vaccination of rhesus macaques with a recombinant measles virus expressing interleukin-12 alters humoral and cellular immune responses. J Infect Dis 188:1553–1561

Huber M, Cattaneo R, Spielhofer P, Orvell C, Norrby E, Messerli M, Perriard JC, Billeter MA (1991) Measles virus phosphoprotein retains the nucleocapsid protein in the cytoplasm. Virology 185:299–308

Janke M, Peeters B, de Leeuw O, Moorman R, Arnold A, Fournier P, Schirrmacher V (2007) Recombinant Newcastle disease virus (NDV) with inserted gene coding for GM-CSF as a new vector for cancer immunogene therapy. Gene Ther 14:1639–1649

Johnson JE, Nasar F, Coleman JW, Price RE, Javadian A, Draper K, Lee M, Reilly PA, Clarke DK, Hendry RM, Udem SA (2007) Neurovirulence properties of recombinant vesicular stomatitis virus vectors in non-human primates. Virology 360:36–49

Karron R, Collins P (2007) Parainfluenza viruses. In: Knipe D, Howley P (eds) Fields virology, vol. 1. Lippincott Williams, Wilkins, Philadelphia, pp 1497–1526

Kato A, Sakai Y, Shioda T, Kondo T, Nakanishi M, Nagai Y (1996) Initiation of Sendai virus multiplication from transfected cDNA or RNA with negative or positive sense. Genes Cells 1:569–579

Klenerman P, Hengartner H, Zinkernagel RM (1997) A non-retroviral RNA virus persists in DNA form. Nature 390:298–301

Kobune F, Sakata H, Sugiura A (1990) Marmoset lymphoblastoid cells as a sensitive host for isolation of measles virus. J Virol 64:700–705

Lamb R, Parks G (2007) Paramyxoviridae: the viruses and their replication. In: Knipe D, Howley P (eds) Fields virology, vol 1. Lippincott Williams Wilkins, Philadelphia, pp 1449–1496

Lawson ND, Stillman EA, Whitt MA, Rose JK (1995) Recombinant vesicular stomatitis viruses from DNA. Proc Natl Acad Sci U S A 92:4477–4481

Li S, Locke E, Bruder J, Clarke D, Doolan DL, Havenga MJ, Hill AV, Liljestrom P, Monath TP, Naim HY, Ockenhouse C, Tang DC, Van Kampen KR, Viret JF, Zavala F, Dubovsky F (2007) Viral vectors for malaria vaccine development. Vaccine 25:2567–2574

Liniger M, Zuniga A, Tamin A, Azzouz-Morin TN, Knuchel M, Marti RR, Wiegand M, Weibel S, Kelvin D, Rota PA, Naim HY (2008) Induction of neutralising antibodies and cellular immune responses against SARS coronavirus by recombinant measles viruses. Vaccine 26:2164–2174

Lorence RM, Roberts MS, O'Neil JD, Groene WS, Miller JA, Mueller SN, Bamat MK (2007) Phase 1 clinical experience using intravenous administration of PV701, an oncolytic Newcastle disease virus. Curr Cancer Drug Targets 7:157–167

Lorin C, Mollet L, Delebecque F, Combredet C, Hurtrel B, Charneau P, Brahic M, Tangy F (2004) A single injection of recombinant measles virus vaccines expressing human immunodeficiency virus (HIV) type 1 clade B envelope glycoproteins induces neutralizing antibodies and cellular immune responses to HIV. J Virol 78:146–157

Luytjes W, Krystal M, Enami M, Parvin JD, Palese P (1989) Amplification, expression, and packaging of foreign gene by influenza virus. Cell 59:1107–1113

Lyles D, Rupprecht C (2007) Rhabdoviridae. In: Knipe D, Howley P (eds) Fields virology, vol 1. Lippincott, Williams, Wilkins, Philadelphia, pp 1363–1408

Martin A, Staeheli P, Schneider U (2006) RNA polymerase II-controlled expression of antigenomic RNA enhances the rescue efficacies of two different members of the *Mononegavirales* independently of the site of viral genome replication. J Virol 80:5708–5715

Monath TP, Guirakhoo F, Nichols R, Yoksan S, Schrader R, Murphy C, Blum P, Woodward S, McCarthy K, Mathis D, Johnson C, Bedford P (2003) Chimeric live, attenuated vaccine against Japanese encephalitis (ChimeriVax-JE): phase 2 clinical trials for safety and immunogenicity, effect of vaccine dose and schedule, and memory response to challenge with inactivated Japanese encephalitis antigen. J Infect Dis 188:1213–1230

Mrkic B, Pavlovic J, Rulicke T, Volpe P, Buchholz CJ, Hourcade D, Atkinson JP, Aguzzi A, Cattaneo R (1998) Measles virus spread and pathogenesis in genetically modified mice. J Virol 72:7420–7427

Naim HY, Ehler E, Billeter MA (2000) Measles virus matrix protein specifies apical virus release and glycoprotein sorting in epithelial cells. EMBO J 19:3576–3585

Nakayama T, Komase K, Uzuka R, Hoshi A, Okafuji T (2001) Leucine at position 278 of the AIK-C measles virus vaccine strain fusion protein is responsible for reduced syncytium formation. J Gen Virol 82:2143–2150

Neumann G, Watanabe T, Ito H, Watanabe S, Goto H, Gao P, Hughes M, Perez DR, Donis R, Hoffmann E, Hobom G, Kawaoka Y (1999) Generation of influenza A viruses entirely from cloned cDNAs. Proc Natl Acad Sci U S A 96:9345–9350

Palin A, Chattopadhyay A, Park S, Delmas G, Suresh R, Senina S, Perlin DS, Rose JK (2007) An optimized vaccine vector based on recombinant vesicular stomatitis virus gives high-level, long-term protection against *Yersinia pestis* challenge. Vaccine 25:741–750

Park KH, Huang T, Correia FF, Krystal M (1991) Rescue of a foreign gene by Sendai virus. Proc Natl Acad Sci U S A 88:5537–5541

Parks CL, Lerch RA, Walpita P, Wang HP, Sidhu MS, Udem SA (2001a) Analysis of the noncoding regions of measles virus strains in the Edmonston vaccine lineage. J Virol 75:921–933

Parks CL, Lerch RA, Walpita P, Wang HP, Sidhu MS, Udem SA (2001b) Comparison of predicted amino acid sequences of measles virus strains in the Edmonston vaccine lineage. J Virol 75:910–920

Patterson JB, Cornu TI, Redwine J, Dales S, Lewicki H, Holz A, Thomas D, Billeter MA, Oldstone MB (2001) Evidence that the hypermutated M protein of a subacute sclerosing panencephalitis measles virus actively contributes to the chronic progressive CNS disease. Virology 291:215–225

Peeters BP, de Leeuw OS, Koch G, Gielkens AL (1999) Rescue of Newcastle disease virus from cloned cDNA: evidence that cleavability of the fusion protein is a major determinant for virulence. J Virol 73:5001–5009

Polo JM, Belli BA, Driver DA, Frolov I, Sherrill S, Hariharan MJ, Townsend K, Perri S, Mento SJ, Jolly DJ, Chang SM, Schlesinger S, Dubensky TW Jr (1999) Stable alphavirus packaging cell lines for Sindbis virus and Semliki Forest virus-derived vectors. Proc Natl Acad Sci U S A 96:4598–4603

Pushko P, Parker M, Ludwig GV, Davis NL, Johnston RE, Smith JF (1997) Replicon-helper systems from attenuated Venezuelan equine encephalitis virus: expression of heterologous genes in vitro and immunization against heterologous pathogens in vivo. Virology 239:389–401

Racaniello VR, Baltimore D (1981) Cloned poliovirus complementary DNA is infectious in mammalian cells. Science 214:916–919

Radecke F, Billeter MA (1996) The nonstructural C protein is not essential for multiplication of Edmonston B strain measles virus in cultured cells. Virology 217:418–421

Radecke F, Spielhofer P, Schneider H, Kaelin K, Huber M, Dotsch C, Christiansen G, Billeter MA (1995) Rescue of measles viruses from cloned DNA. EMBO J 14:5773–5784

Rager M, Vongpunsawad S, Duprex WP, Cattaneo R (2002) Polyploid measles virus with hexameric genome length. EMBO J 21:2364–2372

Roberts A, Rose JK (1999) Redesign and genetic dissection of the rhabdoviruses. Adv Virus Res 53:301–319

Roberts A, Buonocore L, Price R, Forman J, Rose JK (1999) Attenuated vesicular stomatitis viruses as vaccine vectors. J Virol 73:3723–3732

Rose NF, Marx PA, Luckay A, Nixon DF, Moretto WJ, Donahoe SM, Montefiori D, Roberts A, Buonocore L, Rose JK (2001) An effective AIDS vaccine based on live attenuated vesicular stomatitis virus recombinants. Cell 106:539–549

Ryan MD, Drew J (1994) Foot-and-mouth disease virus 2A oligopeptide mediated cleavage of an artificial polyprotein. EMBO J 13:928–933

Schmidt AC, McAuliffe JM, Huang A, Surman SR, Bailly JE, Elkins WR, Collins PL, Murphy BR, Skiadopoulos MH (2000) Bovine parainfluenza virus type 3 (BPIV3) fusion and hemagglutinin-neuraminidase glycoproteins make an important contribution to the restricted replication of BPIV3 in primates. J Virol 74:8922–8929

Schmidt AC, Wenzke DR, McAuliffe JM, St Claire M, Elkins WR, Murphy BR, Collins PL (2002) Mucosal immunization of rhesus monkeys against respiratory syncytial virus subgroups A and B and human parainfluenza virus type 3 by using a live cDNA-derived vaccine based on a host range-attenuated bovine parainfluenza virus type 3 vector backbone. J Virol 76:1089–1099

Schneider H, Kaelin K, Billeter MA (1997) Recombinant measles viruses defective for RNA editing and V protein synthesis are viable in cultured cells. Virology 227:314–322

Schnell MJ, Mebatsion T, Conzelmann KK (1994) Infectious rabies viruses from cloned cDNA. EMBO J 13:4195–4203

Schnell MJ, Foley HD, Siler CA, McGettigan JP, Dietzschold B, Pomerantz RJ (2000) Recombinant rabies virus as potential live-viral vaccines for HIV-1. Proc Natl Acad Sci U S A 97:3544–3549

Schwartz JA, Buonocore L, Roberts A, Suguitan A Jr, Kobasa D, Kobinger G, Feldmann H, Subbarao K, Rose JK (2007) Vesicular stomatitis virus vectors expressing avian influenza H5 HA induce cross-neutralizing antibodies and long-term protection. Virology 366:166–173

Sidhu MS, Chan J, Kaelin K, Spielhofer P, Radecke F, Schneider H, Masurekar M, Dowling PC, Billeter MA, Udem SA (1995) Rescue of synthetic measles virus minireplicons: measles

genomic termini direct efficient expression and propagation of a reporter gene. Virology 208:800–807
Simon ID, Publicover J, Rose JK (2007) Replication and propagation of attenuated vesicular stomatitis virus vectors in vivo: vector spread correlates with induction of immune responses and persistence of genomic RNA. J Virol 81:2078–2082
Singh M, Billeter MA (1999) A recombinant measles virus expressing biologically active human interleukin-12. J Gen Virol 80:101–106
Singh M, Cattaneo R, Billeter MA (1999) A recombinant measles virus expressing hepatitis B virus surface antigen induces humoral immune responses in genetically modified mice. J Virol 73:4823–4828
Skiadopoulos MH, Surman SR, Riggs JM, Orvell C, Collins PL, Murphy BR (2002) Evaluation of the replication and immunogenicity of recombinant human parainfluenza virus type 3 vectors expressing up to three foreign glycoproteins. Virology 297:136–152
Skiadopoulos MH, Schmidt AC, Riggs JM, Surman SR, Elkins WR, St Claire M, Collins PL, Murphy BR (2003a) Determinants of the host range restriction of replication of bovine parainfluenza virus type 3 in rhesus monkeys are polygenic. J Virol 77:1141–1148
Skiadopoulos MH, Vogel L, Riggs JM, Surman SR, Collins PL, Murphy BR (2003b) The genome length of human parainfluenza virus type 2 follows the rule of six, and recombinant viruses recovered from non-polyhexameric-length antigenomic cDNAs contain a biased distribution of correcting mutations. J Virol 77:270–279
Spielhofer P (1995) Generation of standard, variant and chimeric measles viruses from cloned DNA. PhD thesis, University of Zurich, Zurich, Switzerland
Spielhofer P, Bachi T, Fehr T, Christiansen G, Cattaneo R, Kaelin K, Billeter MA, Naim HY (1998) Chimeric measles viruses with a foreign envelope. J Virol 72:2150–2159
Springfeld C, von Messling V, Frenzke M, Ungerechts G, Buchholz CJ, Cattaneo R (2006) Oncolytic efficacy and enhanced safety of measles virus activated by tumor-secreted matrix metalloproteinases. Cancer Res 66:7694–7700
Takeda M, Ohno S, Seki F, Nakatsu Y, Tahara M, Yanagi Y (2005) Long untranslated regions of the measles virus M and F genes control virus replication and cytopathogenicity. J Virol 79:14346–14354
Takeda M, Nakatsu Y, Ohno S, Seki F, Tahara M, Hashiguchi T, Yanagi Y (2006) Generation of measles virus with a segmented RNA genome. J Virol 80:4242–4248
Tangy F, Naim HY (2005) Live attenuated measles vaccine as a potential multivalent pediatric vaccination vector. Viral Immunol 18:317–326
Tani H, Komoda Y, Matsuo E, Suzuki K, Hamamoto I, Yamashita T, Moriishi K, Fujiyama K, Kanto T, Hayashi N, Owsianka A, Patel AH, Whitt MA, Matsuura Y (2007) Replication-competent recombinant vesicular stomatitis virus encoding hepatitis C virus envelope proteins. J Virol 81:8601–8612
Taniguchi T, Palmieri M, Weissmann C (1978) QB DNA-containing hybrid plasmids giving rise to QB phage formation in the bacterial host. Nature 274:223–228
Tatsuo H, Ono N, Tanaka K, Yanagi Y (2000) SLAM (CDw150) is a cellular receptor for measles virus. Nature 406:893–897
Tatsuo H, Ono N, Yanagi Y (2001) Morbilliviruses use signaling lymphocyte activation molecules (CD150) as cellular receptors. J Virol 75:5842–5850
Tubulekas I, Berglund P, Fleeton M, Liljestrom P (1997) Alphavirus expression vectors and their use as recombinant vaccines: a minireview. Gene 190:191–195
Ungerechts G, Springfeld C, Frenzke ME, Lampe J, Johnston PB, Parker WB, Sorscher EJ, Cattaneo R (2007a) Lymphoma chemovirotherapy: CD20-targeted and convertase-armed measles virus can synergize with fludarabine. Cancer Res 67:10939–10947
Ungerechts G, Springfeld C, Frenzke ME, Lampe J, Parker WB, Sorscher EJ, Cattaneo R (2007b) An immunocompetent murine model for oncolysis with an armed and targeted measles virus. Mol Ther 15:1991–1997
Valsamakis A, Schneider H, Auwaerter PG, Kaneshima H, Billeter MA, Griffin DE (1998) Recombinant measles viruses with mutations in the C, V, or F gene have altered growth phenotypes in vivo. J Virol 72:7754–7761

Vigil A, Park MS, Martinez O, Chua MA, Xiao S, Cros JF, Martinez-Sobrido L, Woo SL, Garcia-Sastre A (2007) Use of reverse genetics to enhance the oncolytic properties of Newcastle disease virus. Cancer Res 67:8285–8292

von Messling V, Cattaneo R (2004) Toward novel vaccines and therapies based on negative-strand RNA viruses. Curr Topics Microbiol Immunol 283:281–312

Wang Z, Hangartner L, Cornu TI, Martin LR, Zuniga A, Billeter MA, Naim HY (2001) Recombinant measles viruses expressing heterologous antigens of mumps and simian immunodeficiency viruses. Vaccine 19:2329–2336

Wertz GW, Perepelitsa VP, Ball LA (1998) Gene rearrangement attenuates expression and lethality of a nonsegmented negative strand RNA virus. Proc Natl Acad Sci U S A 95:3501–3506

Whelan SP, Ball LA, Barr JN, Wertz GT (1995) Efficient recovery of infectious vesicular stomatitis virus entirely from cDNA clones. Proc Natl Acad Sci U S A 92:8388–8392

Zuniga A, Wang Z, Liniger M, Hangartner L, Caballero M, Pavlovic J, Wild P, Viret JF, Glueck R, Billeter MA, Naim HY (2007) Attenuated measles virus as a vaccine vector. Vaccine 25:2974–2983

Chapter 8
Measles Virus Interaction with Host Cells and Impact on Innate Immunity

D. Gerlier(✉) and H. Valentin

Contents

Abbreviations.	164
Cellular Receptors and Virus Entry.	165
Natural Cellular Receptors as Partners of MV Glycoproteins.	165
Molecular Changes Associated with Adaptation to Unnatural Host: Use of CD46 as Cellular Receptor by Attenuated Vaccine Strains.	166
The Black Box of Intracellular Viral and Cellular Protein Partnership Leading to Optimal Virus Life Cycle.	168
Partnership in Virus Transcription and Replication.	168
Which Partnership Is Used for Virus Assembly and Budding?.	171
Virus and Cellular Innate Immunity Cross-Talk.	173
Recognition of MV Transcription by RIG-I and Induction of IFN-β.	173
Virus-Induced Syncytium as a Robust Factory for IFN.	175
Anti-innate Immunity and/or Cellular Dysfunction Promoting Properties of P, V and C Proteins.	176
Interplay of MV with Toll-Like Receptors.	178
Modulation of Cellular Functions by N Protein.	179
Cellular Antiviral Effectors Active Against MV.	180
Conclusion and Future Studies.	181
References.	184

Abstract Because viruses are obligate parasites, numerous partnerships between measles virus and cellular molecules can be expected. At the entry level, measles virus uses at least two cellular receptors, CD150 and a yet to be identified epithelial receptor to which the virus H protein binds. This dual receptor strategy illuminates the natural infection and inter-human propagation of this lymphotropic virus. The attenuated vaccine strains use CD46 as an additional receptor, which results in a tropism alteration. Surprisingly, the intracellular viral and cellular

D. Gerlier
Interactions virus cellule-hôte; CNRS, Université de Lyon 1, FRE3011, IFR 62 Laennec, 69372 Lyon Cedex 08, France, e-mail: denis.gerlier@univ-lyon1.fr

protein partnership leading to optimal virus life cycle remains mostly a black box, while the interactions between viral proteins that sustain the RNA-dependant RNA polymerase activity (i.e., transcription and replication), the particle assembly and the polarised virus budding are documented. Hsp72 is the only cellular protein that is known to regulate the virus transcription and replication through its interaction with the viral N protein. The viral P protein is phosphorylated by the casein kinase II with undetermined functional consequences. The cellular partnership that controls the intracellular trafficking of viral components, the assembly and/or the budding of measles virus, remains unknown. The virus to cell innate immunity war is better documented. The 5′ triphosphate-ended virus leader transcript is recognised by RIG-I, a cellular helicase, and induces the interferon response. Measles virus V protein binds to the MDA5 helicase and prevents the MDA5-mediated activation of interferon. By interacting with STAT1 and Jak1, the viral P and V proteins prevent the type I interferon receptor (IFNAR) signalling. The virus N protein interacts with eIF3-p40 to inhibit the translation of cellular mRNA. The H protein binds to TLR2, which then transduces an activation signal and CD150 expression in monocytes. The P protein activates the expression of the ubiquitin modifier A20, thus blocking the TLR4-mediated signalling. Few other partnerships between measles virus components and cellular proteins have been postulated or demonstrated, and they need further investigations to understand their physiopathological outcome.

Abbreviations

MV	Measles virus
(i)(m)DC	(Immature)(mature)dendritic cells
CEF	Chicken embryonic fibroblasts
DI	Defective interfering
Hsp	Heat shock protein
NLS	Nuclear localisation signal
NES	Nuclear export signal
CRM1	Chromosomal region maintenance 1 protein
IFN	Interferon
IFNAR	IFN-α receptor
RNP	Ribonucleoprotein
PNT	P N-terminus
PCT	P C-terminus
VCT	V C-terminus
eIF	Eukaryotic initiation factor
FcγR	Fc gamma receptor
NR	Nucleoprotein receptor
ADAR	Adenosine deaminase
IRF	IFN regulatory factor

Measles virus (MV) is particularly well suited to infect humans, who constitute the unique natural reservoir for this virus. Like all viruses, measles virus is an obligate parasite of a cell. The host cell provides all the necessary substrates, cellular cofactors, protein synthesis, intracellular trafficking and vesicle budding machineries for the synthesis and the assembly of its RNA and protein constituents and for the budding of the virus at the plasma membrane. Multiple constraints have to be dealt with by the virus in order to complete its infection cycle. The target cell has to express an adequate cellular receptor, as well as polymerase cofactors, cytoskeleton cargos, and budding components that can be recruited by viral proteins. The activation and/or maturation of cell hosts results in cellular milieu changes that the virus has to cope with. The virus induced cellular stress and the activation of the cellular innate immunity have to be counteracted or dealt with. Furthermore, the artificial adaptation of measles virus to grow in unnatural host cells can also shed some light on critical molecular parameters involved in the cross-talk between measles virus and the host cell.

The aim of this chapter is to summarise the molecular information dealing with the mutual relationships between MV and host cells during the virus life cycle. Since some aspects are detailed elsewhere in this book, they will only be briefly mentioned here.

Cellular Receptors and Virus Entry

Natural Cellular Receptors as Partners of MV Glycoproteins

MV is primarily a lymphotropic virus that also infects epithelial and neuronal cells. CD150 (also called signalling lymphocytic activation molecule, SLAM) is the cellular receptor that binds to MV H glycoprotein to mediate the virus entry into a large fraction of cells from the immune system, including resting and/or activated B and T lymphocytes, and dendritic cells (DCs) (Condack et al. 2007; de Swart et al. 2007) (see the chapter by Y. Yanagi et al., this volume). It is intriguing that human platelets also expressed CD150 (Nanda et al. 2005)], and, as a consequence, can act as trapping sponge for any virus particle circulating into the blood and/or can bind to any circulating MV-infected cells.

A large body of molecular and cellular evidence (see the chapters by Y. Yanagi et al. and C. Navaratnarajah et al., this volume) predicts the use of an epithelial receptor to mediate MV binding and entry at the basolateral side of polarised epithelial cells. Although neurokinin-1 (NK-1, substance P receptor) seems to promote MV entry into neurons by possibly serving as a receptor for the MV F protein (Makhortova et al. 2007), the molecular pathway leading to MV entry into neuronal cells remains to be better documented (see the chapter by V.A. Young and G. Rall, this volume).

Details on the cellular receptors and their interactions with MV glycoproteins, including structure, binding sites, membrane fusion, intracellular signalling, impact on DC functions and immunosuppression are described in the chapters by Y. Yanagi et al. and C. Navaratnarajah et al., this volume, and by S. Schneider-Schaulies and J. Schneider-Schaulies and B. Hahm, this volume.

Molecular Changes Associated with Adaptation to Unnatural Host: Use of CD46 as Cellular Receptor by Attenuated Vaccine Strains

The forced growth of a virus into a non-natural host cell can be a way to decipher some of the underlying molecular mechanisms which support virus replication in a host cell.

The history of MV started with the isolation of the Edmonston strain, which was isolated in human kidney epithelial cells (Enders and Peebles 1954). A posteriori, since syncytia were observed from the first passage, the virus likely entered the cells by using the postulated epithelial receptor (see the chapters by Y. Yanagi et al. and C. Navaratnarajah et al., this volume). After a few passages in human kidney epithelial cells, the virus was successfully grown either in simian epithelial Vero cells (Rota et al. 1994) or in chicken embryonic fibroblasts (CEFs). It should be noted that the first four passages in CEF were blind, i.e. without cytopathogenic effects (Katz et al. 1958). All vaccine strains currently in use derive from five different type A wild-type MV, including Edmonston, and have been passed several times in embryonated eggs and/or quail embryonic cells and/or most often in CEF (Rota et al. 1994). These MV strains shared four common properties that distinguish them from the currently wild-type MV strains: they belong to the type A strain, they can use human CD46 as a cellular receptor, they grow well in Vero cells and CEF and they are strongly attenuated both in humans and monkeys. Some vaccine strains, such as AIK-C, Zagreb and Moraten, differ in their thymic pathogenicity, as evaluated in a human thymus xenograft model in SCID mice (Valsamakis et al. 2001). Furthermore, successive passages of the Moraten strain in such thymic tissue leads to the reversal from virus attenuation to virulence, although the molecular mechanism has not been investigated (Valsamakis et al. 1999).

Because the MV vaccine strains use both CD46 and CD150 receptors (see the chapters by Y. Yanagi et al., C. Kemper and J.P. Atkinson, and C. Navaratnarajah et al., this volume), they have a tropism alteration when infecting human tonsillar tissue in vitro (Condack et al. 2007). Detailed descriptions of CD46 and its interaction with MV H, including structure, binding sites, membrane fusion, intracellular signalling, impact on macrophage or Treg functions and immunosuppression can be found in the chapters by C. Kemper and J. Atkinson, C. Navaratnarajah et al., S. Schneider-Schaulies and J. Schneider-Schaulies, and B. Hahm, this volume.

When propagated in vitro in their established cell host, the vaccine strains are remarkably stable, with little or no change in sequence with time passages and maintenance in different laboratories all around the world, as well illustrated by the remarkable nucleotide sequence identity of the Schwarz and Moraten strains (Parks et al. 2001a, 2001b; Rota et al. 1994). Unfortunately, the primary genomic sequences of the five parental wild-type strains, as well as those of any other genotype A viruses, are not available and only the sequence of the so-called wild-type Edmonston strain, which has been passed seven times in human kidney epithelial cells and six to eight times in Vero cells, could be determined (Rota et al. 1994). Therefore, it is difficult to associate genome mutations with the adaptation to the avian cell environment and attenuation in vivo. Indeed, there are few mutations in noncoding regions, some of them being common to all Edmonston-derived vaccine strains (Parks et al. 2001a), but their functional impact is unknown. It should be stressed here that an optimal virus life cycle can rely on the tight control of M and F levels by the UTR regions of these two genes (Takeda et al. 2005). The vaccine strains also display changes in coding sequence in all viral proteins, few of them being common to all vaccine strains (Parks et al. 2001b; Rima et al. 1995). One notable common mutation is the N481Y mutation in H, which is absent from the so-called wild-type Edmonston strain and is associated with the usage of CD46 as alternate cellular receptor. Since then, there have been several attempts to adapt currently circulating wild-type MV strain to grow in $CD150^-$ $CD46^+$Vero or $CD150^-$ $CD46^-$ CEF cells searching for the acquisition of common mutations that would sign the adaptation process. Adaptation into $CD150^-$ Vero cells is characterised by initial slow growth of the virus without cytopathic effects for a few passages. At this stage, cell-associated virus is reduced by approximately100-fold and the cell-free virus, which usually accounts for roughly 0.1% of the cell-associated virus, is almost undetectable (Bankamp et al. 2007). After a few passages, the virus production increases with the appearance of cytopathic effects. The associated mutations are usually very limited to a few amino acid changes in one or several of the H, M, N, P/V/C and L proteins. Notably, mutations in H allowing the use of CD46 as cellular receptor can be sporadically observed (Li and Qi 2002; Nielsen et al. 2001; Schneider et al. 2002; Shibahara et al. 1994). However, they are far from appearing systematically (Bankamp et al. 2007; Kouomou and Wild 2002; Shibahara et al. 1994; Takeuchi et al. 2000), although most of the wild-type strains bear the I390, D416 and S446 residues that are additionally required for the efficient use of CD46 (Tahara et al. 2007a). Furthermore, the expression of a wild-type H unable to bind to CD46 in the context of an Edmonston strain allows strong virus growth in Vero cells (Takeuchi et al. 2002), while clinical MV isolates from genotype D3 can enter and replicate into activated splenic lymphocytes from CD46 transgenic mice (Manchester et al. 2000). How a virus with initially poor entry abilities can overcome the entry step to grow to about normal levels in most cells after adaptation remains enigmatic. MV can enter many cells from several species at a very low level by interaction with widely expressed inefficient receptors (Hashimoto et al. 2002). After adaptation without mutations in H and F glycoproteins, it can be speculated that the overall affinity of MV for plasma membranes is increased over a defined threshold level (Hasegawa et al. 2007) because of self–self interaction of some cellular membrane proteins that have been embarked on the virus envelope during the budding process. Further-more, or alternatively, the mutation(s) within the virus replication machinery can compensate for the entry defect. When looking at mutations in other virus genes occurring after adaptation in Vero cells, no mutation was systematically recovered, nor was a single gene systematically found to be mutated, indicating that MV adaptation for optimal growth in Vero cells is multifactorial. Only one recent attempt to adapt a wild-type strain into CEF has been described and it is associated with mutations in internal P/V/C and M proteins (Bankamp et al. 2007). As observed initially (Katz et al. 1958), this adaptation requires few (<10) blind passages before a cytopathic effect is observed. The adaptation to Vero cells does not alleviate the adaptation step for growth in CEF. Adaptation to Vero or CEF resulted in attenuated phenotype in monkeys (Bankamp et al. 2007). The lack of any consensus changes in the sequences of viral proteins, including glycoproteins, after adaptation is even more puzzling, when considering that this adaptation is no longer required for wild-type MV infection of $CD150^+$ Vero cells (Bankamp et al. 2007; Ono et al. 2001) or for the infection of $CD46^+$ CEF by a stain that uses CD46 (Escoffier and Gerlier 1999). The cell host can even be insect cells expressing human CD150 (Liu et al. 2005). However, some animal cells, such as primary lymphocytes from human CD150 or CD46 transgenic mice, remain poorly permissive *in* vitro to the infection by a vaccine MV strain (Evlashev et al. 1999; Ohno et al. 2007), and, in vivo, to wild-type strains, (Ohno et al. 2007). In the latter case, because MV growth is observed when transgenic CD150 mice are also deficient for the type I IFN receptor (IFNAR), the mouse cellular innate immune response is responsible for the stringent control of MeV infection in vivo.

The growth of MV in different host cells can also lead to major changes in the propensity of the virus to generate defective interfering (DI) particles (Whistler et al. 1996). This reflects the efficiency with which the polymerase continues using the same template. Indeed, during MV replication, DI nucleocapsids can be generated by the jump of the polymerase from its template to another one, or further down onto the same template (internal deletion DI) or to the nascent nucleocapsid containing the complementary RNA strand (copyback DI).

The Black Box of Intracellular Viral and Cellular Protein Partnership Leading to Optimal Virus Life Cycle

Partnership in Virus Transcription and Replication

From rare early studies in vitro, cellular factors including tubulin have been reported to be required to allow MeV RNA synthesis (Horikami and Moyer 1995). Likewise, the higher fitness of DCs for MV replication after their maturation supports the idea that a better fitted host cell machinery conditions the virus life cycle (Fugier-Vivier et al. 1997; Murabayashi et al. 2002). Surprisingly, our knowledge of the cellular and viral partnership that drives the virus life cycle has not improved, except for the identification of heat shock protein Hsp72 as a regulator of polymerase activity.

When looking at other members of the *Mononegavirales*, the amount of available information is also rather limited, thus preventing a comprehensive view. The rare either putative or documented cellular partnership are briefly reviewed in the next section.

Virus and Cell Stress Cross-Talk: Heat Shock Proteins (Hsp) as Polymerase Cofactor(s)

MV transcription and replication relies on a complex molecular ballet, the details of which can be found in the chapters by B.K. Rima and W.P. Duprex and S. Longhi, this volume. Briefly, the nucleocapsid made of the genome encapsidated by nucleoprotein N acts as the template of either the P+L transcriptase complex or a N+P+L replicase complex. The association-dissociation cycle of the C-terminal domain of P (XD) with the Box2 (aa 489–506) and Box3 (aa 517–525) regions of the C-terminal domain of N (N_{TAIL}) plays a critical role in allowing the progression of the polymerase along the nucleocapsid template. Notably, the inducible Hsp72 also binds to the same region of N_{TAIL}. The high affinity of Hsp72 binding to Box2 seems to be shared by all MV strains (Zhang et al. 2005), and Hsp72 is proposed to reduce the stability of XD binding to N_{TAIL}, thereby promoting the binding and release cycle that is essential to polymerase processivity (see the chapter by S. Longhi, this volume). The Box3 region of vaccine strains contains an additional low-affinity binding site for Hsp72, which critically relies on the presence of an asparagine at position 522. This renders the virus highly sensitive to the Hsp72 promoting effect on both transcription and replication in vitro, as shown after artificial overexpression of Hsp72 (Carsillo et al. 2006b; Zhang et al. 2005). In vitro, the enhancing effect of Hsp72 overexpression mimics that observed when endogenous Hsp72 is induced by a thermal shock (Vasconcelos et al. 1998a, 1998b). In vivo, hyperthermic preconditioning enhances the clearance of MV from the brain in a mouse model (Carsillo et al. 2004), while the expression of a Hsp72 transgene in mice promotes virus burden and neuropathology (Carsillo et al. 2006a). This opposite outcome of virus replication in the brain of Hsp72 transgenic animals and hyperthermic preconditioned mice indicates that parameters other than induction of Hsp72 control MV replication in the brain.

Hsp90 has been found to act as a chaperone for the L polymerase of several *Mononegavirales*, including *Paramyxoviridae*. Silencing Hsp90 or using specific drug inhibitors results in a decrease in virus replication (Connor et al. 2007). Because of the strong homology of L proteins within the *Mononegavirales* order, it is tempting to speculate that Hsp90 may also be a partner for measles L protein.

Post-translational Modifications of Viral Proteins from the Polymerase Complex: Kinase(s) and Ubiquitin Ligase(s)

P, V and N proteins are phosphorylated on Ser and/or Thr residues (Gombart et al. 1995; Robbins and Bussell 1979; Robbins et al. 1980a, 1980b; Wardrop and Breidis

1991), the latter also being phosphorylated on Tyr residues in persistently infected neuroblastoma cells (Ofir et al. 1996). While soluble N is phosphorylated solely on Ser residues, N in nucleocapsid (N^{NUC}) is also phosphorylated on Thr residues (Gombart et al. 1995). In vitro, P (and likely V) protein is phosphorylated by casein kinase II (CKII) on Ser86, Ser151 and Ser180 residues (Das et al. 1995) and possibly by the zeta isoform of protein kinase C (PKC-ζ) (Liu et al. 1997). The role of these phosphorylation events is unknown, but they are likely critical, as shown for other *Mononegavirales* (Liu et al. 1997) and references therein). Akt, also known as protein kinase B, is a Ser/Thr kinase, which phosphorylates P protein from the parainfluenza virus 5 (PIV5), another *Paramyxoviridae*, and is negatively regulated by PIV5V protein (Sun et al. 2008). Since a specific inhibitor of Akt strongly decreases the release of MV, it is speculated that Akt is also critically involved in the replication of MV. Interestingly, in the absence of MV replication, MV H/ F can inhibit T cell proliferation in vitro (Schlender et al. 1996) and the intracellular pathway involves the impairment of the Akt kinase activation (Avota et al. 2001, 2006). This challenges the possibility of two distinct Akt-dependent molecular mechanisms for infected and uninfected cells, contributing to either MV propagation or immunosuppression.

MV P protein can be ubiquitinated and targeted for degradation by the proteasome, but the ubiquitin ligase involved is unknown (Chen et al. 2005).

The Mysterious Nucleocytoplasmic Shuttling of N and C Proteins and Possible Role of the Nucleus in the Virus Life Cycle

When the N protein is expressed alone, it accumulates in the nucleus, whereas it remains essentially in the cytosol when co-expressed with the P protein, or during the virus infection cycle (Huber et al. 1991; Spehner et al. 1997). Like other *Morbillivirus*, MeV N possesses a leucine-isoleucine-rich nuclear localisation signal (NLS) at position N[70–77] that possesses no similarity to any previously reported NLS. It also bears a leucine-rich nuclear export signal (NES) at position N[425–440], which is independent from the chromosomal region maintenance 1 protein (CRM1) and has no known similarities with other NESs (Sato et al. 2006). Within the soluble, monomeric form of N (N°), P protein binds to the N[4119] region (Bankamp et al. 1996), which overlaps the NLS region of N protein. Thus, by masking the NLS, the P protein may act as a cytosolic retention factor for N. Intranuclear inclusion bodies made of N are consistently found in persistently infected cells (Norrby 1972; Robbins 1983). This implies that N might be shuttling between the cytoplasm and the nucleus during the viral replication steps, and that the nuclear transport of N protein may be involved in the pathogenicity specific to *Morbillivirus*. Alternatively, nucleocytoplasmic shuttling may be required for MV replication since the replication of MV, but not that of the respiratory syncytial virus, is debilitated in enucleated cells (Follett et al. 1976).

When expressed alone, or at the beginning of MV infection, C protein accumulates exclusively in the nucleus because of a classical NLS signal at position

C[41–48]. At a later stage of infection, C protein mainly localises in the cytosol due to a classical CRM1-dependent NES sequence at position C[76–85] (Nishie et al. 2007). Unravelling the mechanism of the nucleocytoplasmic shuttling of C may help decipher some of the functions of this protein.

Mitochondrial Short-Chain Enoyl-CoA Hydratase and MV Replication

Long-term persistent infection of human glioblastoma cell lines with MV is associated with the downmodulation of the mitochondrial short chain enoyl-CoA hydratase (ECHS), which catalyzes the β-oxidation pathway of fatty acid. The silencing of this gene impairs the replication of MV, suggesting a role for the lipid metabolism of the host cells in the virus life cycle (Takahashi et al. 2007).

Which Partnership Is Used for Virus Assembly and Budding?

Virus Assembly and Membrane Rafts

MV assembly occurs at the plasma membrane. Cell membranes are not homogenous. Cholesterol and (glyco)sphingolipids tend to segregate into microdomains called membrane rafts. Membrane rafts act as a scaffold and/or assembly platform for many biological events such as signal transduction and virus assembly (see Chazal and Gerlier 2003, for review). MV assembly occurs within membrane rafts (Fig. 1) (Manie et al. 2000; Vincent et al. 2000). In infected cells, all structural proteins including the mature F_1-F_2 and H proteins are enriched into the raft fraction, while the F_0 precursor remains excluded from the raft fraction. Interestingly, F_0 is cleaved into F_1-F_2 in the trans-Golgi network, where the membrane rafts are formed. The MV F protein is targeted to rafts in the absence of other viral proteins. Because F_0 and H proteins associate in the endothelium reticulum (ER) (Plemper et al. 2001), F can drag H into rafts, whereas H expressed alone is excluded from these microdomains. When the other structural proteins are expressed alone, only the M protein exhibits a low but significant association with membrane rafts. The co-expression of H and F together with M does not result in a significant increase in the amount of M associated with rafts, although M protein interacts with the cytoplasmic tail of F (Spielhofer et al. 1998). The inability of F to drag M into rafts is in agreement with the lack of evidence for co-transport of M with MV glycoproteins for efficient surface targeting (Riedl et al. 2002). Independently from the H and F glycoproteins, the genomic RNA, N, P and L proteins associate with RNPs in the cytosol, allowing the scaffolding of M, which probably targets the M–RNP complex to membrane rafts (Vincent et al. 2000). Indeed, M stability and accumulation at intracellular membranes is a prerequisite for M and nucleocapsid co-transport to the plasma membrane and for subsequent virus assembly and budding (Runkler et al. 2007).

Thus, it is proposed that membrane rafts allow co-localisation and assembly of H and F glycoproteins, on one hand, and of the M–RNP complex, on the other hand. Indeed, detergent-resistant virus-like particles (VLP) budding from infected cells are observed with electron microscopy (Bohn et al. 1986) and the assembly in membrane rafts is functional, since rafts isolated from infected cells are infectious (Manie et al. 2000).

Incidentally, the infection by MV strongly modifies receptor signalling such as CD40 and T cell receptor, which required membrane raft integrity, thus contributing to MV-induced immunosuppression (Avota et al. 2004; Servet-Delprat et al. 1993; Vidalain et al. 2000). Unfortunately, the viral component(s) responsible for these effects have not been identified yet.

Virus Budding

After assembly of MV within membrane rafts, the envelope–RNP complex is ready for budding. After budding, single particles contain several RNPs (Rager et al. 2002). Virus released into cell-free supernatant is partially made of non-raft membranes, with recovery in detergent-resistant membrane rafts of H and F glycoproteins, but of none of the other virus structural proteins (Manie et al. 2000). Thus, either the particles assembled in membrane rafts are not the precursors of budding mature viruses, or after assembly involving the coalescence of several membrane rafts, virus budding through membrane rafts is associated with the capture of adjacent non-raft membranes and simultaneously initiates a shift of the RNPs from raft to non-raft regions. Two observations argue for the latter hypothesis: (a) whereas RNPs are tightly bound to the plasma membrane of infected cells, the RNPs tend to dissociate from the virus envelope after budding (Dubois-Dalcq and Reese 1975) and (b) there is a correlation between a defect in MV budding in a murine cell line (Vincent et al. 1999) and the poor localisation of M protein in membrane rafts from infected cells (Chazal and Gerlier 2003). Virus-like particles have been found to occur upon expression of isolated F or M protein without strong evidence for a synergistic effect (Pohl et al. 2007), but the relevance of these findings to virus budding awaits confirmation. MV assembly and/or budding are dependent on Rab9 since the silencing of endogenous the Rab9 GTPase delays MV release (Murray et al. 2005). Rab9 is involved in the retrotrafficking from recycling endosome and/or late endosome to the trans-Golgi network and regulates the trafficking of lipid raft components. This suggests that proper MV assembly may require the recycling of H and F glycoproteins. That budding is controlled by cellular factors is exemplified by the budding defect of wild-type MV strains in Vero cells and by the rescue of efficient budding after P64S and E89K substitutions in M (Tahara et al. 2005). The effect of these mutations is to strengthen the interaction of M with the cytoplasmic tail of H in a cell type-dependent manner (Tahara et al. 2007b). Likewise, the budding of the AIK-C vaccine strain in monocyte-derived DCs is restricted because of the M instability in these cells (Ohgimoto et al. 2007). The C protein, a product of the P gene, is also required for optimal production of infectious virus (Devaux and Cattaneo

2004). So far, no late domain capable of recruiting the endosomal sorting complex required for transport (ESCRT) machinery has been identified within the M, C protein or any other viral protein.

Topology of Virus Assembly and Budding in Polarised Cells

The topology of MV life cycle in polarised epithelial cells is very specific, with entry at the basolateral side where the postulated epithelial receptor is expressed (V.H.J. Leonard et al. 2008; Tahara et al. 2008) and progeny release from the apical membrane domain to ensure dissemination to other humans (V.H.J. Leonard et al. 2008; see also the chapters by Y. Yanagi et al. and C. Navaratnarajah et al., this volume for details and Fig. 1). The H and F glycoproteins are transported in a non-polarised fashion and to the basolateral membrane domain, respectively (Maisner et al. 1998). The expression of H and F at the basolateral membrane is required for cell-to-cell fusion and MV spreading in vitro and in vivo (Moll et al. 2004). Both basolateral targeting and internalisation of H and F are governed by a tyrosine located in their cytoplasmic tails (Y12 for H, Y549 for F) (Moll et al. 2004). The M protein is responsible for diverting some H and F glycoproteins from the basolateral side to the apical site where the virus buds (Naim et al. 2000). The tyrosine-based targeting motifs of both H and F are also critical for uropod targeting in polarised lymphocytes and for H and F interaction with M (Runkler et al. 2008). The underlying molecular mechanisms are unknown and the recruited cellular factors that ensure the proper intracellular trafficking of all the virus components to the budding site remain to be identified.

Virus and Cellular Innate Immunity Cross-Talk

In vitro (Herschke et al. 2007; Plumet et al. 2007) (and references therein) and in vivo (Devaux et al. 2008) infection by MV activates the innate immunity with the induction of a type I interferon (IFN) response. Several molecular mechanisms that are involved in the dual virus and cellular innate immunity cross-talk have recently been elucidated, including those underlying detection of virus infection by RIG-I and those sustaining the numerous anti-IFN activities exerted by the viral proteins.

Recognition of MV Transcription by RIG-I and Induction of IFN-β

Because RIG-I recognises 5′-triphosphate-ended RNA (Hornung et al. 2006; Pichlmair et al. 2006; Plumet et al. 2007), recognising any (viral) transcription

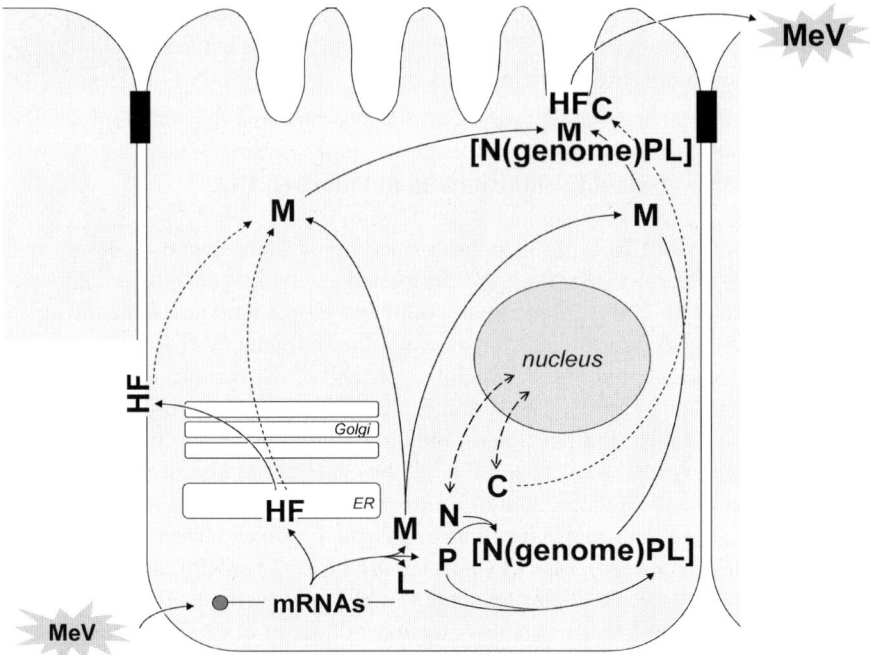

Fig. 1 Trafficking of virus components and MV assembly in polarised epithelial cell. Virus enters at the basolateral site. Among the proteins that are translated from the viral mRNAs in the cytosol, the H and F glycoproteins are targeted to the ER. During replication, the RNA (anti)genome is encapsidated by a helicoidal polymer of N proteins to form the nucleocapsid to which P and L proteins associates. H and F associate in the ER and after maturation into the Golgi they are targeted to the basolateral site. The M protein diverts H/F complexes from the basolateral site and addresses them at the apical site, where it also associates with the nucleocapsid made of the genome, N, P and L proteins. This assembly likely occurs within membrane rafts see Sect. 2.21 and (Chazal and Gerlier 2003 for details). The C protein acts as a budding enhancing factor and then the new virus particle buds from the apical site. *Arrow with broken lines* N and C shuttling between the cytoplasm and the nucleus. The *upper lateral black bars* are the tight junctions between contiguous epithelial cells that delineate the apical and basal sites. *Dotted arrows* indicate alternative or unknown pathway. Note that for the clarity of the scheme, the actual relative position of the ER, Golgi and nucleus is not respected

occurring in the cytosol is proposed (Fig. 2). RIG-I scan cytosolic RNAs, and trigger an IFN response when encountering a free 5′-triphosphate RNA. In other words, a mislocated transcriptional activity is recognised as the hallmark of a foreign invader. During MV infection, the IFN-β gene transcription follows kinetics that parallels that of the virus transcription. This is due to the recognition of 5′-triphosphate-ended leader (and/or trailer) RNA by RIG-I. The leader synthesised both in vitro and in vivo, is efficient in activating the IFN-β response provided that it is delivered into the cytosol as a 5′-trisphosphate-ended RNA. The genome and antigenome are the only two other viral RNA species that are 5′-triphosphate-ended, but their tight encapsidation into N protein prevents them from activating RIG-I

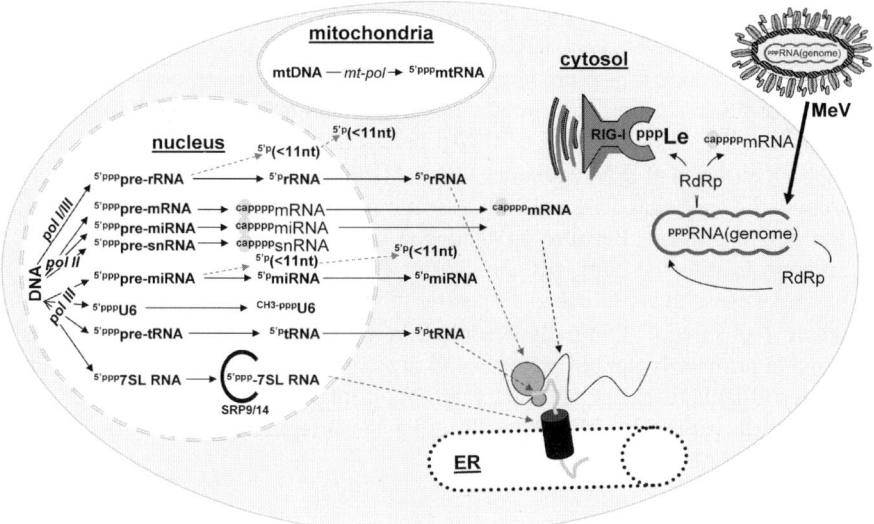

Fig. 2 Selective recognition of 5' triphosphate-ended virus leader transcript by RIG-I. All cellular RNAs but one are either cleaved or modified at their 5' end before migrating from the nucleus into the cytosol. The 7SL RNA is the only cell RNA which remains 5' triphosphate-ended, but this extremity is likely shielded by the protein components of the translocon that associate with this RNA in the nucleolus (see Plumet et al. 2007 for details). Because of their tight association with the N protein assembled into the nucleocapsid, the virus genome and antigenome are also shielded away from RIG-I recognition, and the viral mRNAs are capped. Abbreviations: *mt* mitochondrial, *pol* (cellular) DNA-dependent RNA polymerase, *r* ribosomal, *m* messenger, *mi* micro, *t* transfer, *SRP* signal recognition peptide, *Le* (MV) leader RNA, *RdRp* (MV) RNA-dependent RNA polymerase

(Plumet et al. 2007). Then, during the co-evolution between MV and its cellular host, which selective pressure is necessary for maintaining such short leader transcript? Is this the price to be paid by the polymerase to find the first gene start? The induction of IFN can also be enhanced by the presence of DI particles (Shingai et al. 2007). In this case, the generation of double-stranded RNA could be responsible, at least in part, for the activation of the RIG-dependant IFN response (Strahle et al. 2006, 2007).

Virus-Induced Syncytium as a Robust Factory for IFN

MV infection is characterised by the formation of multinuclear giant cells (MGCs) or syncytia that time-lapse microscopy studies revealed to have a highly dynamic behaviour and an unexpectedly long lifespan in vitro (Herschke et al. 2007). In addition, MGCs derived from human epithelial cells or mature conventional dendritic cells (mDCs) are much higher IFN-α/β producers than single cells (Herschke et al. 2007). Both fusion and IFN-β response amplification is inhibited in a dose-dependent way by a fusion inhibitory peptide after MV infection of epithelial cells. Furthermore, following the cell–cell fusion, RIG-I and IFN-β gene deficiencies can be trans-complemented to induce IFN-β production. The enhancing effect of cell–cell fusion on the IFN response requires infectious virus and occurs at low and high multiplicities of infection. This amplification of IFN-β production is associated with a sustained nuclear localisation of IFN regulatory factor 3 (IRF-3) in MV-induced MGCs derived from both the epithelial cells and mDCs, while IRF-7 upregulation is poorly sensitive to the fusion process. Because MV-induced cell–cell fusion amplifies type I IFN production in infected cells, MGCs should significantly contribute to the antiviral immune response in vivo. Indeed, a specific subset of infected MGCs called Warthin-Finkeldey cells (WFCs), initially described in infants who died from acute measles, is found in primary lymphoid organs, such as the thymus, and in the germinal centres and interfollicular areas of secondary lymphoid organs, where activated B, T, and dendritic cells expressing CD150 reside and where adaptive immune response occurs (de Swart et al. 2007; Gerlier et al. 2006).

Anti-innate Immunity and/or Cellular Dysfunction Promoting Properties of P, V and C Proteins

Blockade of IFNAR Signalling by Virus P, V and C Proteins

To circumvent the IFN-induced antiviral states, the MV has developed redundant strategies that block IFN activation and/or IFNAR signal transduction. The MV V protein interacts with MDA5, which belongs to the RIG-like receptors devoted to detecting anomalous RNA in the cytosol and to activating the type I IFN response (Andrejeva et al. 2004; Childs et al. 2007). Binding occurs through the interaction of the C-terminal domain of V (VCT) with the helicase domain of MDA5 and results in the inhibition of the IFN response mediated by MDA5.

The N-terminus of the P protein (PNT), which is common to both of the P and V proteins, binds to STAT1, with the Tyr110 residue playing a key role in this interaction (Caignard et al. 2007; Devaux et al. 2007; Ohno et al. 2004). As a consequence, both P and V proteins can inhibit the phosphorylation of STAT1 and the signal cascade downstream to the IFNAR (Caignard et al. 2007; Devaux et al. 2007; Takeuchi et al. 2003; Yokota et al. 2003). In addition, the V protein binds to the kinase domain JH1 of Jak1. Although the primary binding domain on V is located on the common PNT domain, its C-terminal VCT domain seems to be required to stabilise the interaction (Caignard et al. 2007). V protein makes trimeric complexes with Jak1 and STAT1 and is co-immunoprecipitated within larger complexes associated with IFNAR signalling (Palosaari et al. 2003; Yokota et al. 2003). Abrogation of V and P binding to STAT1 is not sufficient to prevent the inhibition of STAT1 phosphorylation by Jak1, indicating that binding to either STAT1 or Jak1 can result in inhibitory activity. Interestingly, the CAM vaccine strain exhibits an Y110C substitution (Fontana et al. 2008) that could contribute to its attenuated phenotype.

It should be stressed that, in contrast to V proteins from mumps and other Rubulaviruses that strongly bind to the UV-damaged DNA binding (DBB1) protein (see (Chen and Gerlier 2006) for review), MV V protein does not act as a ubiquitin E3 ligase subunit for STAT degradation.

The nonstructural C protein can inhibit IFNAR downstream signalling (Shaffer et al. 2003), but this effect is weak and variable depending on the MV strain, (Fontana et al. 2008). In infected cells, the C protein has been found to be associated in complex with IFNAR, STAT1 and the scaffolding protein RACK1 (Yokota et al. 2003), but the underlying inhibitory mechanism is unknown. In a host cell with fully competent IFNAR signalling, a recombinant MV lacking the C protein does not grow well because of a reduced virus protein translation and replication. This is associated with the phosphorylation of the eIF2α translation factor (Nakatsu et al. 2006), likely by the protein kinase R (PKR), one of the antiviral effectors activated by the IFN response. How can the C protein prevent the phosphorylation of eIF2α remains to be unraveled.

In summary, P, V, and possibly C, proteins counteract the IFN-α/β-induced antiviral activity through a number of molecular strategies. This diversity may explain the variation in the ability of MeV to counteract the IFN response according to the cell host and virus strain combinations.

V and C Protein as Virulence Factors with Anti-inflammatory Properties

In vivo, MeV V and C proteins function as virulence factors (Devaux et al. 2008; Patterson et al. 2000; Valsamakis et al. 1998). While the Edmonston tag strain, expressing a P and V protein deficient in IFN-antagonistic activity (Ohno et al. 2004), kills CD46-transgenic mice, V-deficient Edmonston tag hardly replicates after intracranial inoculation of 1-day-old suckling mice, and C-deficient MV replicates but fails to kill the animals (Patterson et al. 2000). Accordingly, both of the V- and C-deficient Edmonston tag also display an attenuated phenotype in IFNAR-null mice expressing human CD46 (Mrkic et al. 2000), as does a Vero-adapted wild-type MV strain with four mutations in P/V/C genes in IFNAR-null mice expressing human CD150 (Druelle et al. 2008). Likewise, C- or V-deficient wild-type viruses (Devaux et al. 2008; Takeuchi et al. 2005) are attenuated in monkeys.

Besides the anti-IFN-α/β properties of both V and C proteins in vivo (Devaux et al. 2008), both C and V proteins are required for MV to strongly inhibit the inflammatory response in the peripheral blood monocytic cells of infected monkeys. In the absence of V or C protein, virus-induced TNFα downregulation is alleviated and IL-6 is upregulated (Devaux et al. 2008).

In vitro, C-deficient Edmonston tag MV, but not V-deficient Edmonston tag, shows reduced growth in human peripheral blood monocytes and in human thymus tissues xenografted into SCID mice (Escoffier et al. 1999; Valsamakis et al. 1998), whereas the former virus replicates efficiently in Vero cells and in murine neurons (Patterson et al. 2000; Radecke and Billeter 1996). This cell-type-specific requirement for the C protein indicates that it probably interacts with lymphoid cell-type-specific factors yet to be determined.

Finally, since both V and C proteins can also modulate the virus polymerase activity in a minigenome assay (Parks et al. 2006; Reutter et al. 2001; Witko et al. 2006), the dysregulation of the viral RNA synthesis resulting from the lack of these proteins may contribute to hampering the virus dissemination.

In conclusion, in vivo, MV C and V proteins encode virulence functions that likely operate via multiple and separate mechanisms.

Inhibition of Pirh2 Degradation by P Protein

The human p53-induced-RING-H2 (Pirh2) is a ubiquitin E3 ligase belonging to the RING-like family. Through its C-terminus and RING domain, Pirh2 interacts with the XD domain of P and, together with MV N, can form a ternary complex. MV P efficiently stabilises PIRH2 expression and prevents its ubiquitination (Chen et al. 2005). Our current knowledge on PIRH2 substrates and functions is rather limited. It is a p53-inducible E3 ligase involved in the ubiquitination of p53 and the ε-subunit of the coatmer complex, ε–COP, which is part of the COP-I secretion complex. As such, PIRH2 can regulate cell proliferation and the secretion machinery (Maruyama et al. 2008). Interestingly, the virulence factor ORF3 protein of the porcine circovirus also interacts with PIRH2 (Liu et al. 2007). Additional work is needed to delineate the physiopathological relevance of the interaction of MV P with PIRH2.

Inhibition of p73-Dependent Activation by V Protein

MV V protein was found to bind to the DNA-binding domains of both p53 and p73, members of the p53 family. While it does not affect p53-mediated gene activation, the V protein shows partial inhibition of a p73-dependent gene activation. As such, it has been suggested that V protein could act as an inhibitor of cell death by preventing the activation of the apoptosis regulator PUMA (Cruz et al. 2006).

Interplay of MV with Toll-Like Receptors

Wild-type, but not vaccine, MV strains activate cells via both human and murine TLR2, a property of the H protein. TLR2 activation by MV H protein induces surface expression of CD150 and the secretion of pro-inflammatory cytokines such as IL-6 in human monocytic cells and macrophages. The unique property of MV wild-type strains of activating TLR2-dependent signals could contribute not only to immune activation, but also to viral spread and pathogenicity by allowing monocyte infection (Bieback et al. 2002).

TLR7/8-mediated IFN-α/β induction requires internalisation in an acidic compartment in which genomic RNA can be uncoated from N protein. Some MV is probably endocytosed because MHC-II restricted presentation of N protein is observed (Gerlier et al. 1994). Whether MV genomic RNA becomes accessible for binding to TLR7/8 to activate the type I IFN response remains to be determined. MV infection of human plasmacytoid dendritic cells (pDCs), has been reported to inhibit the IFN response induced by ligand binding to TLR7 or TLR9 (Schlender et al. 2005). However, this effect seems to be virus strain-specific (Druelle et al. 2008).

The MV P protein has transactivating properties due to the acid-rich PNT domain (Chen et al. 2003), and the ubiquitin-modifying enzyme A20 gene is directly activated by the MV P protein (Yokota et al. 2008). P indirectly interacts with a negative regulatory motif in the A20 gene promoter and releases the suppression of A20 transcription in monocytes, but not in epithelial cells. The upregulation of A20 blocks the lipopolysaccharide (LPS)-induced TLR4-mediated activation of NF-κ B and AP-1 in monocytic cells. The suppression of A20 expression by siRNA restored LPS-induced signalling in infected cells. This cell-type-specific suppression of the inflammatory response may contribute to the evasion of MV from the host immune system. However, how P reaches the nucleus is unknown. One possibility is that it could use the N protein as a nucleocytoplasmic shuttle (see Sect. 2.13).

Modulation of Cellular Functions by N Protein

Inhibition of Cellular Translation by the N Protein

N protein binds to the eukaryotic initiation factor 3 (eIF3-p40) (Sato et al. 2007). In vitro, the interaction between MV N and eIF3-p40 inhibits the translation of a reporter mRNAs in a rabbit reticulocyte lysate translation system in a dose-dependent manner. This effect is reminiscent of the observation made with the closely related Sendai virus, where the addition of nucleocapsid templates inhibits the translation of a reporter mRNA (Pelet et al. 2005). In vivo, the inducible expression of MV N inhibits the synthesis of a transfected reporter protein, as well as overall protein synthesis. This is a first observation indicating that MV can repress the host cell translation. However, MV N does not inhibit the translation of an RNA containing the intergenic region of Plautia stali intestine virus, which can assemble 80S ribosomes in the absence of canonical initiation factors. Could it be that viral mRNAs selectively escape requirement for eIF3-p40 initiation factor for translation of viral proteins?

Activation of IRF-3 by N Protein

MV N can activate IRF-3 and thereby induce CCL5 (also called RANTES), a proinflammatory cytokine, but not IFN-β (ten Oever et al. 2002). After MV infection, IRF-3 is phosphorylated at the key Ser385 and Ser386 residues. Activation of IRF-3, which required active MV transcription, is also mimicked by the transient expression of the N protein. Moreover, IRF-3 and a cellular kinase could be co-immunoprecipitated with N. The IRF-3 binding region and N binding region map to N_{TAIL}[415–523] and residues IRF-3[198–394], respectively (ten Oever et al. 2002). However, the direct interaction between N and IRF-3 has been recently challenged by using a panel of biochemical and biophysical techniques and by the lack of any evidence for co-localisation of N and IRF-3 in human cells (Colombo et al., unpublished observations).

Extracellular N Protein Binds to FcγRII and NR Receptors

The MV N is an internal protein within infected cells and virus. However, MV N is also released in the extracellular milieu by dying cells (Laine et al. 2003). In addition, the N protein seems to be able to translocate from the cytosol into the late endocytic compartment from which it can be excreted when the cells express the FcγRII receptor (Marie et al. 2004). To date, three cell surface receptors can bind MV N: specific B cell receptor (BCR) (i.e. the cell surface immunoglobulin) expressed on B lymphocytes (Graves et al. 1984) low-affinity Fc receptor for IgG of type II (FcγRII, CD32) Ravanel et al. 1997), and an as yet unidentified protein receptor called nucleoprotein receptor (NR) (Laine et al. 2003, 2005). The region of MV N responsible for binding to human FcγRIIB maps to N_{CORE}, likely through an exposed loop (aa 122–144) (Laine et al. 2005). NR is constitutively expressed on fibroblasts, epithelial, lymphoid and myeloid cells. NR is not detected on resting T cells, but upregulated upon activation (Laine et al. 2003). The region of MV N responsible for binding to NR maps to the Box1 (aa 401–420) region of N_{TAIL} (Laine et al. 2005). Whether NR is a ubiquitous and then unique molecule or a family of related receptors is unknown. Since MV N binds to FcγRIIB and NR via N_{CORE} and N_{TAIL}, respectively, MV N can simultaneously bind both to FcγRIIB and NR on the same cell type such as B cells, monocytes/macrophages and DCs. In contrast, MV N interacts only with NR on human activated T cells (Laine et al. 2003). In vitro, binding of MV N to FcγRIIB and NR induces apoptosis and inhibits both spontaneous and induced cell proliferation, respectively (Laine et al. 2003, 2005; Ravanel et al. 1997). It also inhibits the immunoglobulin (Ig) synthesis of CD40- or BCR-activated human B lymphocytes in the presence of IL-2 and IL-10 (Ravanel et al. 1997). From in vivo studies in mice, MV N binding to FcγRIIB appears to contribute to the cellular immunosuppression induced by MV (Marie et al. 2004) (see the chapter by S. Schneider-Schaulies and J. Schneider-Schaulies, this Volume).

Cellular Antiviral Effectors Active Against MV

MxA as Cell-Type-Specific Inhibitor of MV Growth

MxA is induced by IFN and can inhibit MV replication. MxA belongs to the class of dynamin-like large guanosine triphosphatases (GTPases) known to be involved in intracellular vesicle trafficking and organelle homeostasis. It accumulates in the cytoplasm and is partly associated with a COP-I-positive sub-compartment of the ER. (Haller et al. 2007). Stable expression in human glioblastoma (Schneider-Schaulies et al. 1994) leads to a 100-fold decrease in virus transcription, whereas stable expression in monocytic cell lines inhibits only the expression of the H and F glycoproteins at a post-transcriptional level (Schnorr et al. 1993). The underlying mechanism is unknown.

ADAR as a Predicted Antiviral Effector

Natural infection of neuronal tissue in vivo is often associated with virus defective in assembly, with mutations or deletions of the M protein, and the cytoplasmic tails of H and/or F glycoproteins (Cattaneo and Rose 1993; Cattaneo et al. 1986). Most of these mutations result from biased U to C hypermutations (Cattaneo et al. 1989). The candidate proposed for this activity is the cellular RNA-specific adenosine deaminase (ADAR) that deaminates adenosine into inosine. As a consequence, during the following replication cycle, a guanosine is added as a complementary base to the inosine. ADAR is an IFN-inducible RNA-editing enzyme acting on double-stranded RNA (Hoopengardner 2006). Because during RNA synthesis the genome RNA remained wrapped into the nucleocapsid (see the chapters by B.K. Rima and W.P. Duprex and S. Longhi, this volume), a collapse transcription model has been proposed. A nascent transcript may occasionally remain base-paired with the genome in the close vicinity of an active polymerase and become the target of unwinding and modification activity by cellular enzymes including ADAR (Bass et al. 1989; Cattaneo and Billeter 1992; see also the chapters by V.A. Young and G. Rall and M.B. Oldstone, this Volume).

Conclusion and Future Studies

A few more that 20 partnerships between measles viral encoded molecules and cellular proteins have been identified or postulated so far, all listed in Table 1. This certainly represents only a tiny part of the highly complex partnership that the virus establishes with its host cells.

Over the last 15 years, there have been many attempts to elucidate the molecular mechanism governing the entry of MV into host cells. So far, the data have revolutionised our view on the virus infection cycle in humans. MV primarily infects CD150-expressing cells of the immune system, which is possible through the epithelial barrier because of the presence of periscopic dendritic cells (Hammand and Lambrecht 2008), and secondarily infects peripheral epithelial cells by using a basolateral epithelial receptor that should be soon identified. Because of the apical release of the virus, this disseminates the virus from person to person by exhaled droplets. The contribution of the neo-acquisition of the CD46 receptor usage by vaccine strains to their attenuated phenotype with fairly strong immunogenicity remains unclear.

Table 1 Known and hypothetical cell and virus molecular partnerships

Viral molecule	Cellular protein	Binding parameters	Function	Selected references
N	Hsp72	N_{TAIL} Box2 [489–506]	Competition with P XD to enhance the polymerase processivity?	Zhang et al. 2002, 2005
		K_D = ~20nM N_{TAIL} Box3 [505–525 & N522] K_D = ~2μM	Increase transcription and replication	Carsillo et al. 2006b
	?	N[425–440] ⇔ ?	Nuclear localisation signal	Sato et al. 2006
	?	N[70–77] ⇔ ?	Nuclear export signal	
	(Ser/Thr kinase)	Ser and Thr phosphorylation at unknown sites	Phosphorylation of Thr restricted to N^{NUC}	Gombart et al. 1995
	eIF3-p40	N[81–192] ⇔ eIF3-p40	Inhibition of translation of cellular mRNA	Sato et al. 2007
	Fcγ RII	N_{CORE} [122–144 ?]	Inhibition of B and DC functions and induction of apoptosis	Laine et al. 2005; Ravanel et al. 1997
	Postulated NR	N_{TAIL} Box. 1 [401–420]	Inhibition of cell proliferation	Laine et al. 2003, 2005
P	Casein kinase II	P [S86, S151, S180] ⇔ CKII	Regulation of polymerase activity?	Das et al. 1995
	STAT1	P [Y110] ⇔ STAT1	Prevents phosphorylation of STAT1 and thus IFNAR signalling	Devaux et al. 2007
	A20 gene promoter	PNT ⇔ Negative regulatory motif	Activation of A20 expression that leads to blockade of TLR4-mediated transduction signalling	Yokota et al. 2008
	PIRH2	P XD ⇔ PIRH2 C-term	Stabilisation of PIRH2 by inhibition of (auto?) ubiquitination	Chen et al. 2003
V	MDA5 (RIG-like receptor)	VCT ⇔ MDA5 helicase	Inhibition of MDA5-mediated activation of IFN-β	Andrejeva et al. 2004; Childs et al. 2007
	STAT1	PNT (Tyr 110) ⇔ STAT1	Prevents phosphorylation of STAT1 and thus IFNAR signalling	Caignard et al. 2007; Ohno et al. 2004
	Jak1	PNT (+VCT) ⇔ Jak1 JH1 kinase domain		Caignard et al. 2007
	p73	V ⇔ p73 DNA binding domain	Inhibition of p73-dependent gene activation	Cruz et al. 2006
C	?	C[41–48] ⇔ ?	Nuclear localisation signal	Nishie et al. 2004
	CMR1	C[76–85] ⇔ CMR1	Nuclear export signal	

Table 1 (continued)

Viral molecule	Cellular protein	Binding parameters	Function	Selected references
H	CD150 (or SLAM)	H[β 5 sheet, I194] ⇔ CD150 V domain K_D = ~100 nM	Virus binding and entry into cells of the immune system	Hashiguchi et al. 2007; Navaratnarajah et al. 2008; Tatsuo et al. 2000
	Postulated epithelial receptor	H[L482, F483, P497, Y541, Y543] ⇔ ?	Virus binding and entry at the basolateral site of polarized epithelial cells	V.H.J. Leonard et al.; Tahara et al. 2008
	CD46 (or MCP)	H[β 3 to β 6 sheet, Y481, G546] ⇔ CD46[SCR1+2] K_D = ~100 nM	Virus binding and entry (vaccine strains only)	Buchholz et al. 1997; Li and Qi 1998; Naniche et al. 1993; Navaratnarajah et al. 2008
	TLR2	H ectodomain ⇔ TLR2 ectodomain	Induction of CD150 expression on monocytes	Bieback et al. 2002
Leader RNA	RIG-I	5'triphosphate end ⇔ RIG RD	Induction of IFN-β response	Cui et al. 2008; Plumet et al. 2007
Genome RNA	ADAR (?)	dsRNA ⇔ ADAR ?	Biased hypermutation U→C	Bass et al. 1989; Cattaneo et al. 1988

The RIG-I recognition of the cytosolic transcription by MV has been identified as the molecular mechanism responsible for the activation of the IFN-β response, and several pieces of the large molecular puzzle underlying the virus to innate cellular response have been unravelled, notably with the discovery of the inhibition of the Jak1/STAT1 transduction pathway and MDA5-dependent IFN activation by P and/or V proteins. Some molecular interactions have been identified, but their physiopathological role needs to be unveiled. In contrast, Hsp72 is so far the only cellular factor identified as a regulator of the transcription and replication process. Although the viral molecular interplay leading to virus assembly, packaging into the virus envelope and polarised budding has been largely investigated, the cellular machinery used by the virus for the intracellular trafficking of the viral components and enabling their proper assembly and the budding process remains undefined. Few MV proteins can bind to cell-surface proteins and transduces a signal that can disturb the functions of cells from the immune system. Last, the molecular mechanism(s) allowing the adaptation of MV so that they grow in heterologous host cells remains a mystery.

Acknowledgements The authors thank R. Cattaneo, Y. Yanagi, B. Hahm, P. Rota and S. Longhi for having shared their chapter contents, unpublished data and/or for their useful comments, critical readings and/or discussion.

References

Andrejeva J, Childs KS, Young DF, Carlos TS, Stock N, Goodbourn S, Randall RE (2004) The V proteins of paramyxoviruses bind the IFN-inducible RNA helicase, mda-5, and inhibit its activation of the IFN-beta promoter. Proc Natl Acad Sci U S A 101:17264–17269

Avota E, Avots A, Niewiesk S, Kane LP, Bommhardt U, ter Meulen V, Schneider-Schaulies S (2001) Disruption of Akt kinase activation is important for immunosuppression induced by measles virus. Nat Med 7:725–731

Avota E, Muller N, Klett M, Schneider-Schaulies S (2004) Measles virus interacts with and alters signal transduction in T-cell lipid rafts. J Virol 78:9552–9559

Avota E, Harms H, Schneider-Schaulies S (2006) Measles virus induces expression of SIP110, a constitutively membrane clustered lipid phosphatase, which inhibits T cell proliferation. Cell Microbiol 8:1826–1839

Bankamp B, Horikami SM, Thompson PD, Huber M, Billeter M, Moyer SA (1996) Domains of the measles virus N protein required for binding to P protein and self-assembly. Virology 216:272–277

Bankamp B, Hodge G, McChesney MB, Bellini WJ, Rota PA (2007) Genetic changes that affect the virulence of measles virus in a rhesus macaque model. Virology 373:39–50

Bass BL, Weintraub H, Cattaneo R, Billeter MA (1989) Biased hypermutation of viral RNA genomes could be due to unwinding/modification of double-stranded RNA. Cell 56:331

Bieback K, Lien E, Klagge IM, Avota E, Schneider-Schaulies J, Duprex WP, Wagner H, Kirschning CJ, Ter Meulen V, Schneider-Schaulies S (2002) Hemagglutinin protein of wild-type measles virus activates toll-like receptor 2 signaling. J Virol 76:8729–8736

Bohn W, Rutter G, Hohenberg H, Mannweiler K, Nobis P (1986) Involvement of actin filaments in budding of measles virus: studies on cytoskeletons of infected cells. Virology 149:91–106

Buchholz CJ, Koller D, Devaux P, Mumenthaler C, Schneider-Schaulies J, Braun W, Gerlier D, Cattaneo R (1997) Mapping of the primary binding site of measles virus to its receptor CD46. J Biol Chem 272:22072–22079

Caignard G, Guerbois M, Labernardiere JL, Jacob Y, Jones LM, Wild F, Tangy F, Vidalain PO (2007) Measles virus V protein blocks Jak1-mediated phosphorylation of STAT1 to escape IFN-alpha/beta signaling. Virology 368:351–362

Carsillo T, Carsillo M, Niewiesk S, Vasconcelos D, Oglesbee M (2004) Hyperthermic pre-conditioning promotes measles virus clearance from brain in a mouse model of persistent infection. Brain Res 1004:73–82

Carsillo T, Traylor Z, Choi C, Niewiesk S, Oglesbee M (2006a) hsp72, a host determinant of measles virus neurovirulence. J Virol 80:11031–11039

Carsillo T, Zhang X, Vasconcelos D, Niewiesk S, Oglesbee M (2006b) A single codon in the nucleocapsid protein C terminus contributes to in vitro and in vivo fitness of Edmonston measles virus. J Virol 80:2904–2912

Cattaneo R, Billeter MA (1992) Mutations and A/I hypermutations in measles virus persistent infections. Curr Top Microbiol Immunol 176:63–74

Cattaneo R, Rose JK (1993) Cell fusion by the envelope glycoproteins of persistent measles viruses which caused lethal human brain disease. J Virol 67:1493–1502

Cattaneo R, Schmid A, Rebmann G, Baczko K, Ter Meulen V, Bellini WJ, Rozenblatt S, Billeter MA (1986) Accumulated measles virus mutations in a case of subacute sclerosing panencephalitis: interrupted matrix protein reading frame and transcription alteration. Virology 154:97–107

Cattaneo R, Schmid A, Eschle D, Baczko K, ter Meulen V, Billeter MA (1988) Biased hypermutation and other genetic changes in defective measles viruses in human brain infections. Cell 55:255–265

Cattaneo R, Kaelin K, Baczko K, Billeter MA (1989) Measles virus editing provides an additional cysteine-rich protein. Cell 56:759–764

Chazal N, Gerlier D (2003) Virus entry, assembly, budding, and membrane rafts. Microbiol Mol Biol Rev 67:226–237

Chen M, Gerlier D (2006) Viral hijacking of cellular ubiquitination pathways as an anti-innate immunity strategy. Viral Immunol 19:349–362

Chen M, Cortay JC, Gerlier D (2003) Measles virus protein interactions in yeast: new findings and caveats. Virus Res 98:123–129

Chen M, Cortay JC, Logan IR, Sapountzi V, Robson CN, Gerlier D (2005) Inhibition of ubiquitination and stabilization of human ubiquitin E3 ligase PIRH2 by measles virus phosphoprotein. J Virol 79:11824–11836

Childs K, Stock N, Ross C, Andrejeva J, Hilton L, Skinner M, Randall R, Goodbourn S (2007) mda-5, but not RIG-I, is a common target for paramyxovirus V proteins. Virology 359:190–200

Condack C, Grivel JC, Devaux P, Margolis L, Cattaneo R (2007) Measles virus vaccine attenuation: suboptimal infection of lymphatic tissue and tropism alteration. J Infect Dis 196:541–549

Connor JH, McKenzie MO, Parks GD, Lyles DS (2007) Antiviral activity and RNA polymerase degradation following Hsp90 inhibition in a range of negative strand viruses. Virology 362:109–119

Cruz CD, Palosaari H, Parisien JP, Devaux P, Cattaneo R, Ouchi T, Horvath CM (2006) Measles virus V protein inhibits p53 family member p73. J Virol 80:5644–5650

Cui S, Eisenacher K, Kirchhofer A, Brzozka K, Lammens A, Lammens K, Fujita T, Conzelmann KK, Krug A, Hopfner KP (2008) The C-terminal regulatory domain is the RNA 5'-triphosphate sensor of RIG-I. Mol Cell 29:169–179

Das T, Schuster A, Schneider-Schaulies S, Banerjee AK (1995) Involvement of cellular casein kinase II in the phosphorylation of measles virus P protein: identification of phosphorylation sites. Virology 211:218–226

de Swart RL, Ludlow M, de Witte L, Yanagi Y, van Amerongen G, McQuaid S, Yuksel S, Geijtenbeek TB, Duprex WP, Osterhaus AD (2007) Predominant infection of CD150+ lymphocytes and dendritic cells during measles virus infection of macaques. PLoS Pathog 3:e178

Devaux P, Cattaneo R (2004) Measles virus phosphoprotein gene products: conformational flexibility of the P/V protein amino-terminal domain and C protein infectivity factor function. J Virol 78:11632–11640

Devaux P, von Messling V, Songsungthong W, Springfeld C, Cattaneo R (2007) Tyrosine 110 in the measles virus phosphoprotein is required to block STAT1 phosphorylation. Virology 360:72–83

Devaux P, Hodge G, McChesney MB, Cattaneo R (2008) Attenuation of V- or C-defective measles viruses: infection control by the inflammatory and interferon responses of rhesus monkeys. J Virol 82:5359–5367

Druelle J, Sellin CI, Waku-Kouomou D, Horvat B, Wild FT (2008) Wild type measles virus attenuation independent of type I IFN. Virol J 5:22

Dubois-Dalcq M, Reese TS (1975) Structural changes in the membrane of vero cells infected with a paramyxovirus. J Cell Biol 67:551–565

Enders JF, Peebles TC (1954) Propagation in tissue cultures of cytopathogenic agents from patients with measles. Proc Soc Exp Biol Med 86:277–286

Escoffier C, Gerlier D (1999) Infection of chicken embryonic fibroblasts by measles virus: adaptation at the virus entry level. J Virol 73:5220–5224

Escoffier C, Manie S, Vincent S, Muller CP, Billeter M, Gerlier D (1999) Nonstructural C protein is required for efficient measles virus replication in human peripheral blood cells. J Virol 73:1695–1698

Evlashev A, Valentin H, Rivailler P, Azocar O, Rabourdin-Combe C, Horvat B (2001) Differential permissivity to measles virus infection of human and CD46-transgenic murine lymphocytes. J Gen Virol 82:2125–2129

Follett EA, Pringle CR, Pennington TH (1976) Events following the infections of enucleate cells with measles virus. J Gen Virol 32:163–175

Fontana JM, Bankamp B, Bellini WJ, Rota PA (2008) Regulation of interferon signaling by the C, V proteins from attenuated and wild-type strains of measles virus. Virology 374:71–81

Fugier-Vivier I, Servet-Delprat C, Rivailler P, Rissoan MC, Liu YJ, Rabourdin-Combe C (1997) Measles virus suppresses cell-mediated immunity by interfering with the survival and functions of dendritic and T cells. J Exp Med 186:813–823

Gerlier D, Trescol-Biemont MC, Varior-Krishnan G, Naniche D, Fugier-Vivier I, Rabourdin-Combe C (1994) Efficient major histocompatibility complex class II-restricted presentation of measles virus relies on hemagglutinin-mediated targeting to its cellular receptor human CD46 expressed by murine B cells. J Exp Med 179:353–358

Gerlier D, Valentin H, Laine D, Rabourdin-Combe C, Servet-Delprat C (2006) Subversion of the immune system by measles virus: a model for the intricate interplay between a virus and the human immune system. In: Lachman PG, Oldstone MBA (eds) Microbial subversion of host immunity. Caister Academic, Norwalk, UK, pp 225–292

Gombart AF, Hirano A, Wong TC (1995) Nucleoprotein phosphorylated on both serine and threonine is preferentially assembled into the nucleocapsids of measles virus. Virus Res 37:63–73

Graves M, Griffin DE, Johnson RT, Hirsch RL, de Soriano IL, Roedenbeck S, Vaisberg A (1984) Development of antibody to measles virus polypeptides during complicated and uncomplicated measles virus infections. J Virol 49:409–412

Haller O, Staeheli P, Kochs G (2007) Interferon-induced Mx proteins in antiviral host defense. Biochimie 89:812–818

Hammad H, Lambrecht BN (2008) Dendritic cells and epithelial cells: linking innate and adaptive immunity in asthma. Nat Rev Immunol 8:193–204

Hasegawa K, Hu C, Nakamura T, Marks JD, Russell SJ, Peng KW (2007) Affinity thresholds for membrane fusion triggering by viral glycoproteins. J Virol 81:13149–13157

Hashiguchi T, Kajikawa M, Maita N, Takeda M, Kuroki K, Sasaki K, Kohda D, Yanagi Y, Maenaka K (2007) Crystal structure of measles virus hemagglutinin provides insight into effective vaccines. Proc Natl Acad Sci U S A 104:19535–19540

Hashimoto K, Ono N, Tatsuo H, Minagawa H, Takeda M, Takeuchi K, Yanagi Y (2002) SLAM (CD150)-independent measles virus entry as revealed by recombinant virus expressing green fluorescent protein. J Virol 76:6743–6749

Herschke F, Plumet S, Duhen T, Azocar O, Druelle J, Laine D, Wild TF, Rabourdin-Combe C, Gerlier D, Valentin H (2007) Cell-cell fusion induced by measles virus amplifies the type I interferon response. J Virol 81:12859–12871

Hoopengardner B (2006) Adenosine-to-inosine RNA editing: perspectives and predictions. Mini Rev Med Chem 6:1213–1216

Horikami SM, Moyer SA (1995) Structure, transcription, and replication of measles virus. Curr Top Microbiol Immunol 191:35–50

Hornung V, Ellegast J, Kim S, Brzozka K, Jung A, Kato H, Poeck H, Akira S, Conzelmann KK, Schlee M, Endres S, Hartmann G (2006) 5'-Triphosphate RNA is the ligand for RIG-I. Science 314:994–997

Huber M, Cattaneo R, Spielhofer P, Orvell C, Norrby E, Messerli M, Perriard JC, Billeter MA (1991) Measles virus phosphoprotein retains the nucleocapsid protein in the cytoplasm. Virology 185:299–308

Katz SL, Milovanovic MV, Enders JF (1958) Propagation of measles virus in cultures of chick embryo cells. Proc Soc Exp Biol Med 97:23–29

Kouomou DW, Wild TF (2002) Adaptation of wild-type measles virus to tissue culture. J Virol 76:1505–1509

Laine D, Trescol-Biemont MC, Longhi S, Libeau G, Marie JC, Vidalain PO, Azocar O, Diallo A, Canard B, Rabourdin-Combe C, Valentin H (2003) Measles virus (MV) nucleoprotein binds to a novel cell surface receptor distinct from FcgammaRII via its C-terminal domain: role in MV-induced immunosuppression. J Virol 77:11332–11346

Laine D, Bourhis JM, Longhi S, Flacher M, Cassard L, Canard B, Sautes-Fridman C, Rabourdin-Combe C, Valentin H (2005) Measles virus nucleoprotein induces cell-proliferation arrest and apoptosis through NTAIL-NR, NCORE-FcgammaRIIB1 interactions, respectively. J Gen Virol 86:1771–1784

Leonard VH, Sinn PL, Hodge G, Miest T, Devaux P, Oezguen N, Braun W, McCray PB Jr, McChesney MB, Cattaneo R (2008) Measles virus blind to its epithelial cell receptor remains virulent in rhesus monkeys but cannot cross the airway epithelium and is not shed. J Clin Invest 118:2448–2458

Li L, Qi Y (2002) A novel amino acid position in hemagglutinin glycoprotein of measles virus is responsible for hemadsorption and CD46 binding. Arch Virol 147:775–786

Li LY, Qi YP (1998) The point mutations in hemagglutinin gene of measles virus are responsible for alteration in hemadsorption. Sheng Wu Hua Xue Yu Sheng Wu Wu Li Xue Bao (Shanghai) 30:488–494

Liu J, Zhu Y, Chen I, Lau J, He F, Lau A, Wang Z, Karuppannan AK, Kwang J (2007) The ORF3 protein of porcine circovirus type 2 interacts with porcine ubiquitin E3 ligase Pirh2 and facilitates p53 expression in viral infection. J Virol 81:9560–9567

Liu X, Zhou W, Zhang P, Xu Q, Hu C, Chen X, Yao L, Li L, Qi Y (2005) Expression of SLAM (CDw150) on Sf9 cell surface using recombinant baculovirus mediates measles virus infection in the nonpermissive cells. Microbes Infect 7:1235–1245

Liu Z, Huntley CC, De BP, Das T, Banerjee AK, Oglesbee MJ (1997) Phosphorylation of canine distemper virus P protein by protein kinase C-zeta and casein kinase II. Virology 232:198–206

Maisner A, Klenk H, Herrler G (1998) Polarized budding of measles virus is not determined by viral surface glycoproteins. J Virol 72:5276–5278

Makhortova NR, Askovich P, Patterson CE, Gechman LA, Gerard NP, Rall GF (2007) Neurokinin-1 enables measles virus trans-synaptic spread in neurons. Virology 362:235–244

Manchester M, Eto DS, Valsamakis A, Liton PB, Fernandez-Munoz R, Rota PA, Bellini WJ, Forthal DN, Oldstone MB (2000) Clinical isolates of measles virus use CD46 as a cellular receptor. J Virol 74:3967–3974

Manie SN, Debreyne S, Vincent S, Gerlier D (2000) Measles virus structural components are enriched into lipid raft microdomains: a potential cellular location for virus assembly. J Virol 74:305–311

Marie JC, Saltel F, Escola JM, Jurdic P, Wild TF, Horvat B (2004) Cell surface delivery of the measles virus nucleoprotein: a viral strategy to induce immunosuppression. J Virol 78:11952–11961

Maruyama S, Miyajima N, Bohgaki M, Tsukiyama T, Shigemura M, Nonomura K, Hatakeyama S (2008) Ubiquitylation of epsilon-COP by PIRH2 and regulation of the secretion of PSA. Mol Cell Biochem 307:73–82

Miyajima N, Takeda M, Tashiro M, Hashimoto K, Yanagi Y, Nagata K, Takeuchi K (2004) Cell tropism of wild-type measles virus is affected by amino acid substitutions in the P, V, M proteins, or by a truncation in the C protein. J Gen Virol 85:3001–3006

Moll M, Klenk HD, Herrler G, Maisner A (2001) A single amino acid change in the cytoplasmic domains of measles virus glycoproteins H, F alters targeting, endocytosis, and cell fusion in polarized Madin-Darby canine kidney cells. J Biol Chem 276:17887–17894

Moll M, Pfeuffer J, Klenk HD, Niewiesk S, Maisner A (2004) Polarized glycoprotein targeting affects the spread of measles virus in vitro and in vivo. J Gen Virol 85:1019–1027

Mrkic B, Odermatt B, Klein MA, Billeter MA, Pavlovic J, Cattaneo R (2000) Lymphatic dissemination and comparative pathology of recombinant measles viruses in genetically modified mice. J Virol 74:1364–1372

Murabayashi N, Kurita-Taniguchi M, Ayata M, Matsumoto M, Ogura H, Seya T (2002) Susceptibility of human dendritic cells (DCs) to measles virus (MV) depends on their activation stages in conjunction with the level of CDw150: role of Toll stimulators in DC maturation and MV amplification. Microbes Infect 4:785–794

Murray JL, Mavrakis M, McDonald NJ, Yilla M, Sheng J, Bellini WJ, Zhao L, Le Doux JM, Shaw MW, Luo CC, Lippincott-Schwartz J, Sanchez A, Rubin DH, Hodge TW (2005) Rab9 GTPase is required for replication of human immunodeficiency virus type 1, filoviruses, and measles virus. J Virol 79:11742–11751

Naim HY, Ehler E, Billeter MA (2000) Measles virus matrix protein specifies apical virus release and glycoprotein sorting in epithelial cells. EMBO J 19:3576–3585

Nakatsu Y, Takeda M, Ohno S, Koga R, Yanagi Y (2006) Translational inhibition and increased interferon induction in cells infected with C protein-deficient measles virus. J Virol 80:11861–11867

Nanda N, Andre P, Bao M, Clauser K, Deguzman F, Howie D, Conley PB, Terhorst C, Phillips DR (2005) Platelet aggregation induces platelet aggregate stability via SLAM family receptor signaling. Blood 106:3028–3034

Naniche D, Varior-Krishnan G, Cervoni F, Wild TF, Rossi B, Rabourdin-Combe C, Gerlier D (1993) Human membrane cofactor protein (CD46) acts as a cellular receptor for measles virus. J Virol 67:6025–6032

Navaratnarajah CK, Vongpunsawad S, Oezguen N, Stehle T, Braun W, Hashiguchi T, Maenaka K, Yanagi Y, Cattaneo R (2008) Dynamic interaction of the measles virus hemagglutinin with its receptor SLAM. J Biol Chem 283:11763–11771

Nielsen L, Blixenkrone-Moller M, Thylstrup M, Hansen NJ, Bolt G (2001) Adaptation of wild-type measles virus to CD46 receptor usage. Arch Virol 146:197–208

Nishie T, Nagata K, Takeuchi K (2007) The C protein of wild-type measles virus has the ability to shuttle between the nucleus and the cytoplasm. Microbes Infect 9:344–354

Norrby E (1972) Intracellular accumulation of measles virus nucleocapsid and envelope antigens. Microbios 5:31–40

Ofir R, Weinstein Y, Bazarsky E, Blagerman S, Wolfson M, Hunter T, Rager-Zisman B (1996) Tyrosine phosphorylation of measles virus P-phosphoprotein in persistently infected neuroblastoma cells. Virus Genes 13:203–210

Ohgimoto K, Ohgimoto S, Ihara T, Mizuta H, Ishido S, Ayata M, Ogura H, Hotta H (2007) Difference in production of infectious wild-type measles and vaccine viruses in monocyte-derived dendritic cells. Virus Res 123:1–8

Ohno S, Ono N, Takeda M, Takeuchi K, Yanagi Y (2004) Dissection of measles virus V protein in relation to its ability to block alpha/beta interferon signal transduction. J Gen Virol 85:2991–2999

Ohno S, Ono N, Seki F, Takeda M, Kura S, Tsuzuki T, Yanagi Y (2007) Measles virus infection of SLAM (CD150) knockin mice reproduces tropism and immunosuppression in human infection. J Virol 81:1650–1659

Ono N, Tatsuo H, Hidaka Y, Aoki T, Minagawa H, Yanagi Y (2001) Measles viruses on throat swabs from measles patients use signaling lymphocytic activation molecule (CDw150) but not CD46 as a cellular receptor. J Virol 75:4399–4401

Palosaari H, Parisien JP, Rodriguez JJ, Ulane CM, Horvath CM (2003) STAT protein interference and suppression of cytokine signal transduction by measles virus V protein. J Virol 77:7635–7644

Parks CL, Lerch RA, Walpita P, Wang HP, Sidhu MS, Udem SA (2001a) Analysis of the noncoding regions of measles virus strains in the Edmonston vaccine lineage. J Virol 75:921–933

Parks CL, Lerch RA, Walpita P, Wang HP, Sidhu MS, Udem SA (2001b) Comparison of predicted amino acid sequences of measles virus strains in the Edmonston vaccine lineage. J Virol 75:910–920

Parks CL, Witko SE, Kotash C, Lin SL, Sidhu MS, Udem SA (2006) Role of V protein RNA binding in inhibition of measles virus minigenome replication. Virology 348:96–106

Patterson JB, Thomas D, Lewicki H, Billeter MA, Oldstone MB (2000) V, C proteins of measles virus function as virulence factors in vivo. Virology 267:80–89

Pelet T, Miazza V, Mottet G, Roux L (2005) High throughput screening assay for negative single stranded RNA virus polymerase inhibitors. J Virol Methods 128:29–36

Pichlmair A, Schulz O, Tan CP, Naslund TI, Liljestrom P, Weber F, Reis e Sousa C (2006) RIG-I-mediated antiviral responses to single-stranded RNA bearing 5'-phosphates. Science 314:997–1001

Plemper RK, Hammond AL, Cattaneo R (2001) Measles virus envelope glycoproteins hetero-oligomerize in the endoplasmic reticulum. J Biol Chem 276:44239–44246

Plumet S, Herschke F, Bourhis JM, Valentin H, Longhi S, Gerlier D (2007) Cytosolic 5'-triphosphate ended viral leader transcript of measles virus as activator of the RIG I-mediated interferon response. PLoS ONE 2:e279

Pohl C, Duprex WP, Krohne G, Rima BK, Schneider-Schaulies S (2007) Measles virus M, F proteins associate with detergent-resistant membrane fractions and promote formation of virus-like particles. J Gen Virol 88:1243–1250

Radecke F, Billeter MA (1996) The nonstructural C protein is not essential for multiplication of Edmonston B strain measles virus in cultured cells. Virology 217:418–421

Rager M, Vongpunsawad S, Duprex WP, Cattaneo R (2002) Polyploid measles virus with hexameric genome length. EMBO J 21:2364–2372

Ravanel K, Castelle C, Defrance T, Wild TF, Charron D, Lotteau V, Rabourdin-Combe C (1997) Measles virus nucleocapsid protein binds to FcgammaRII and inhibits human B cell antibody production. J Exp Med 186:269–278

Reutter GL, Cortese-Grogan C, Wilson J, Moyer SA (2001) Mutations in the measles virus C protein that up regulate viral RNA synthesis. Virology 285:100–109

Riedl P, Moll M, Klenk HD, Maisner A (2002) Measles virus matrix protein is not cotransported with the viral glycoproteins but requires virus infection for efficient surface targeting. Virus Res 83:1–12

Rima BK, Earle JA, Baczko K, Rota PA, Bellini WJ (1995) Measles virus strain variations. Curr Top Microbiol Immunol 191:65–83

Robbins SJ (1983) Progressive invasion of cell nuclei by measles virus in persistently infected human cells. J Gen Virol 64:2335–2338

Robbins SJ, Bussell RH (1979) Structural phosphoproteins associated with purified measles virions and cytoplasmic nucleocapsids. Intervirology 12:96–102

Robbins SJ, Bussell RH, Rapp F (1980a) Isolation and partial characterization of two forms of cytoplasmic nucleocapsids from measles virus-infected cells. J Gen Virol 47:301–310

Robbins SJ, Fenimore JA, Bussell RH (1980b) Structural phosphoproteins associated with measles virus nucleocapsids from persistently infected cells. J Gen Virol 48:445–449

Rota JS, Wang ZD, Rota PA, Bellini WJ (1994) Comparison of sequences of the H, F, and N coding genes of measles virus vaccine strains. Virus Res 31:317–330

Runkler N, Pohl C, Schneider-Schaulies S, Klenk HD, Maisner A (2007) Measles virus nucleocapsid transport to the plasma membrane requires stable expression and surface accumulation of the viral matrix protein. Cell Microbiol 9:1203–1214

Runkler N, Dietzel E, Moll M, Klenk HD, Maisner A (2008) Glycoprotein targeting signals influence the distribution of measles virus envelope proteins and virus spread in lymphocytes. J Gen Virol 89:687–696

Sato H, Masuda M, Miura R, Yoneda M, Kai C (2006) Morbillivirus nucleoprotein possesses a novel nuclear localization signal and a CRM1-independent nuclear export signal. Virology 352:121–130

Sato H, Masuda M, Kanai M, Tsukiyama-Kohara K, Yoneda M, Kai C (2007) Measles virus N protein inhibits host translation by binding to eIF3-p40. J Virol 81:11569–11576

Schlender J, Schnorr JJ, Spielhoffer P, Cathomen T, Cattaneo R, Billeter MA, ter Meulen V, Schneider-Schaulies S (1996) Interaction of measles virus glycoproteins with the surface of uninfected peripheral blood lymphocytes induces immunosuppression in vitro. Proc Natl Acad Sci U S A 93:13194–13199

Schlender J, Hornung V, Finke S, Gunthner-Biller M, Marozin S, Brzozka K, Moghim S, Endres S, Hartmann G, Conzelmann KK (2005) Inhibition of toll-like receptor 7- and 9-mediated alpha/beta interferon production in human plasmacytoid dendritic cells by respiratory syncytial virus and measles virus. J Virol 79:5507–5515

Schneider U, von Messling V, Devaux P, Cattaneo R (2002) Efficiency of measles virus entry and dissemination through different receptors. J Virol 76:7460–7467

Schneider-Schaulies S, Schneider-Schaulies J, Schuster A, Bayer M, Pavlovic J, ter Meulen V (1994) Cell type-specific MxA-mediated inhibition of measles virus transcription in human brain cells. J Virol 68:6910–6917

Schnorr JJ, Schneider-Schaulies S, Simon-Jodicke A, Pavlovic J, Horisberger MA, ter Meulen V (1993) MxA-dependent inhibition of measles virus glycoprotein synthesis in a stably transfected human monocytic cell line. J Virol 67:4760–4768

Servet-Delprat C, Vidalain PO, Bausinger H, Manie S, Le Deist F, Azocar O, Hanau D, Fischer A, Rabourdin-Combe C (2000) Measles virus induces abnormal differentiation of CD40 ligand-activated human dendritic cells. J Immunol 164:1753–1760

Shaffer JA, Bellini WJ, Rota PA (2003) The C protein of measles virus inhibits the type I interferon response. Virology 315:389–397

Shibahara K, Hotta H, Katayama Y, Homma M (1994) Increased binding activity of measles virus to monkey red blood cells after long-term passage in Vero cell cultures. J Gen Virol 75:3511–3516

Shingai M, Ebihara T, Begum NA, Kato A, Honma T, Matsumoto K, Saito H, Ogura H, Matsumoto M, Seya T (2007) Differential type I IFN-inducing abilities of wild-type versus vaccine strains of measles virus. J Immunol 179:6123–6133

Spehner D, Drillien R, Howley PM (1997) The assembly of the measles virus nucleoprotein into nucleocapsid-like particles is modulated by the phosphoprotein. Virology 232:260–268

Spielhofer P, Bachi T, Fehr T, Christiansen G, Cattaneo R, Kaelin K, Billeter MA, Naim HY (1998) Chimeric measles viruses with a foreign envelope. J Virol 72:2150–2159

Strahle L, Garcin D, Kolakofsky D (2006) Sendai virus defective-interfering genomes and the activation of interferon-beta. Virology 351:101–111

Strahle L, Marq JB, Brini A, Hausmann S, Kolakofsky D, Garcin D (2007) Activation of the beta interferon promoter by unnatural Sendai virus infection requires RIG-I and is inhibited by viral C proteins. J Virol 81:12227–12237

Sun M, Fuentes SM, Timani K, Sun D, Murphy C, Lin Y, August A, Teng MN, He B (2008) Akt plays a critical role in replication of nonsegmented negative-stranded RNA viruses. J Virol 82:105–114

Tahara M, Takeda M, Yanagi Y (2005) Contributions of matrix and large protein genes of the measles virus Edmonston strain to growth in cultured cells as revealed by recombinant viruses. J Virol 79:15218–15225

Tahara M, Takeda M, Seki F, Hashiguchi T, Yanagi Y (2007a) Multiple amino acid substitutions in hemagglutinin are necessary for wild-type measles virus to acquire the ability to use receptor CD46 efficiently. J Virol 81:2564–2572

Tahara M, Takeda M, Yanagi Y (2007b) Altered interaction of the matrix protein with the cytoplasmic tail of hemagglutinin modulates measles virus growth by affecting virus assembly and cell-cell fusion. J Virol 81:6827–6836

Tahara M, Takeda M, Shirogane Y, Hashiguchi T, Ohno S, Yanagi Y (2008) Measles virus infects both polarized epithelial and immune cells using distinctive receptor-binding sites on its hemagglutinin. J Virol 82:4630–4637

Takahashi M, Watari E, Shinya E, Shimizu T, Takahashi H (2007) Suppression of virus replication via down-modulation of mitochondrial short chain enoyl-CoA hydratase in human glioblastoma cells. Antiviral Res 75:152–158

Takeda M, Ohno S, Seki F, Nakatsu Y, Tahara M, Yanagi Y (2005) Long untranslated regions of the measles virus M, F genes control virus replication and cytopathogenicity. J Virol 79:14346–14354

Takeuchi K, Miyajima N, Kobune F, Tashiro M (2000) Comparative nucleotide sequence analyses of the entire genomes of B95a cell-isolated and vero cell-isolated measles viruses from the same patient. Virus Genes 20:253–257

Takeuchi K, Takeda M, Miyajima N, Kobune F, Tanabayashi K, Tashiro M (2002) Recombinant wild-type and Edmonston strain measles viruses bearing heterologous H proteins: role of H protein in cell fusion and host cell specificity. J Virol 76:4891–4900

Takeuchi K, Kadota SI, Takeda M, Miyajima N, Nagata K (2003) Measles virus V protein blocks interferon (IFN)-alpha/beta but not IFN-gamma signaling by inhibiting STAT1 and STAT2 phosphorylation. FEBS Lett 545:177–182

Takeuchi K, Takeda M, Miyajima N, Ami Y, Nagata N, Suzaki Y, Shahnewaz J, Kadota S, Nagata K (2005) Stringent requirement for the C protein of wild-type measles virus for growth both in vitro and in macaques. J Virol 79:7838–7844

Tatsuo H, Ono N, Tanaka K, Yanagi Y (2000) SLAM (CDw150) is a cellular receptor for measles virus. Nature 406:893–897

tenOever BR, Servant MJ, Grandvaux N, Lin R, Hiscott J (2002) Recognition of the measles virus nucleocapsid as a mechanism of IRF-3 activation. J Virol 76:3659–3669

Valsamakis A, Schneider H, Auwaerter PG, Kaneshima H, Billeter MA, Griffin DE (1998) Recombinant measles viruses with mutations in the C, V, or F gene have altered growth phenotypes in vivo. J Virol 72:7754–7761

Valsamakis A, Auwaerter PG, Rima BK, Kaneshima H, Griffin DE (1999) Altered virulence of vaccine strains of measles virus after prolonged replication in human tissue. J Virol 73:8791–8797

Valsamakis A, Kaneshima H, Griffin DE (2001) Strains of measles vaccine differ in their ability to replicate in an damage human thymus. J Infect Dis 183:498–502

Vasconcelos DY, Cai XH, Oglesbee MJ (1998a) Constitutive overexpression of the major inducible 70 kDa heat shock protein mediates large plaque formation by measles virus. J Gen Virol 79:2239–2247

Vasconcelos D, Norrby E, Oglesbee M (1998b) The cellular stress response increases measles virus-induced cytopathic effect. J Gen Virol 79:1769–1773

Vidalain PO, Azocar O, Servet-Delprat C, Rabourdin-Combe C, Gerlier D, Manie S (2000) CD40 signaling in human dendritic cells is initiated within membrane rafts. EMBO J 19:3304–3313

Vincent S, Spehner D, Manie S, Delorme R, Drillien R, Gerlier D (1999) Inefficient measles virus budding in murine L.CD46 fibroblasts. Virology 265:185–195

Vincent S, Gerlier D, Manie SN (2000) Measles virus assembly within membrane rafts. J Virol 74:9911–9915

Wardrop EA, Briedis DJ (1991) Characterization of V protein in measles virus-infected cells. J Virol 65:3421–3428

Whistler T, Bellini WJ, Rota PA (1996) Generation of defective interfering particles by two vaccine strains of measles virus. Virology 220:480–484

Witko SE, Kotash C, Sidhu MS, Udem SA, Parks CL (2006) Inhibition of measles virus minireplicon-encoded reporter gene expression by V protein. Virology 348:107–119

Yokota S, Saito H, Kubota T, Yokosawa N, Amano K, Fujii N (2003) Measles virus suppresses interferon-alpha signaling pathway: suppression of Jak1 phosphorylation and association of viral accessory proteins, C, V, with interferon-alpha receptor complex. Virology 306:135–146

Yokota S, Okabayashi T, Yokosawa N, Fujii N (2008) Measles virus P protein suppresses Toll-like receptor signal through up-regulation of ubiquitin-modifying enzyme A20. FASEB J 22:74–83

Zhang X, Glendening C, Linke H, Parks CL, Brooks C, Udem SA, Oglesbee M (2002) Identification and characterization of a regulatory domain on the carboxyl terminus of the measles virus nucleocapsid protein. J Virol 76:8737–8746

Zhang X, Bourhis JM, Longhi S, Carsillo T, Buccellato M, Morin B, Canard B, Oglesbee M (2005) Hsp72 recognizes a P binding motif in the measles virus N protein C-terminus. Virology 337:162–174

Index

A
A and B boxes, 93
A box, 86
Actin, 81
ADAR, 181
Additional transcription units (ATUs), 140, 141, 147
Alphaviruses, 132, 150
Atypical measles, 8

B
B95a, 14, 15, 19, 21
B box, 86
Biased U to C hypermutations, 181
Borna virus (BV), 138
Burkitt's lymphoma, 24

C
Canine distemper virus (CDV), 21, 22, 24
Cap-dependent scanning, 84, 94
Cap structure, 87
CD46, 14–17, 19–23, 60, 65–69, 139, 140, 145, 147, 152, 166–168, 177, 181
CD150, 15, 165–168, 176–178, 181
CD46 transgenic mice, 167, 168, 177
CDV. *See* Canine distemper virus (CDV)
Cell entry, 59–70
Cellular innate immunity, 165, 173
Cellular stress, 165
Cellular translation, 179
Central conserved region (CCR), 109–110
Chick cell virus, 6
Chicken embryonic fibroblasts, 166
Chimera, 136
Clades, 83
Copies, 84, 90, 91, 94
Copy-choice transcription, 131, 149
Copy numbers, 91, 94
Co-transcriptional editing, 88, 89

C protein, 81, 82, 84, 85, 88, 94, 95, 170–174, 176–178
Crystal structure, 19, 22, 25
Cytoskeleton, 79, 81

D
DC-SIGN, 20, 24
Defective interfering (DIs) particles, 78, 91, 93, 134, 149, 168
Designated receptors, 70–71
Disorder-promoting residues, 112
Dissemination, 60
Dynamics of RNA Accumulation, 89, 91

E
Editing, 83, 88, 89, 93
Edmonston, David, 5
Edmonston Zagreb MV strain, 144
Electron microscopy, 110
Electron paramagnetic resonance (EPR) spectroscopy, 114
Embryonated hen's eggs, 5
Encapsidation signal, 106
Enders, J.F., 4–6, 8
Enders philosophy, 7
Epithelial cell receptor, 20, 22–25, 60, 165, 166, 173, 181
EpR, 60, 61, 65, 67–69
Eukaryotic initiation factor 3, 179
Expanded Program on Immunization (EPI), 9

F
FcγRII, 182
F glycoproteins, 167, 171–174, 181
Fidelity, 94
Flaviviruses, 132, 146
Fusion activation, 64–68, 70, 72
Fusion protein, 63–64

193

G

Genotypes, 83
Glialfibrillary-acidic protein, 81
Glycoprotein complex, 60–62, 64, 65
Gradient, 78
Gradient of gene expression, 90, 92

H

Heat-shock protein Hsp72, 121, 168, 169, 183
Helper viruses, 134
Hemagglutinin, 60, 62, 63
Hepatitis B virus (HBV), 144–145
Herringbone structure, 79
Hilleman, Maurice, 8
HIV. *See* Human immunodeficiency virus (HIV)
Hodgkin's disease, 24
H proteins, 171
Human amnion cells, 5
Human immunodeficiency virus (HIV), 24
Human immunoglobulin, 7
Human kidneys, 4–6
Human papilloma virus (HPV), 144
Hypermutation, 92, 139

I

IDPs, 107–108
Ifnarko/CD46Ge mice, 144–146
IL-6, 177, 178
IL-12, 142
Importations, 9
Inactivated vaccine, 8
Induced folding, 108
Inflammatory response, 177, 179
Influenza viruses, 132–134, 141, 144
In situ hybridisation, 81
Interferon, 173
Interferon type I, 139, 144
Intergenic (Ig) sequences, 81, 89
In vitro transcription system, 84
IRF-3, 176, 179, 180

J

Jak1, 176, 183

L

Leader, 84, 87, 93, 174, 175
3' Leader, 87
L protein, 79, 81, 84–87, 91, 92

M

Malaria, 144, 150, 151
Marker proteins, 142
MDA5, 176, 183
Measles Partnership (Measles Initiative), 9
Membrane rafts, 171, 172, 174
Midireplicons, 135
Minireplicons, 133–135, 140
Misincorporation, 149
Molecular Recognition Element, 112
Monkeys, 5, 6
Mononegavirales, 130, 132–134, 136–138, 150, 152–155
Morbillivirus, 14, 21, 22, 138
MoREs, 112
Morley, D.C., 8
M protein, 81, 92, 95, 168, 171–174, 181
Multivalent vaccines, 130, 143, 150, 151
MVA-T7, 136
MxA, 180
MxA protein, 92, 94

N

N°, 106, 120
Natively unfolded proteins, 108
N CORE, 108
Neomycin (Neo) resistance, 135
Neurokinin-1 (NK-1), 20
Neuronal cells, 92
Nigeria, 8, 9
N NUC, 106
N NUC–P–L, 105, 106
N NUC ûP complex, 117
N°–P complex, 106, 107, 109, 111
N–P–L tripartite complex, 118
N Protein, 169, 170, 174, 179, 180
N-RNA complexes, 105, 119
N TAIL, 107
Nuclear export signal (NES), 106
Nuclear localization signal (NLS), 106
Nucleocapsidlike particles, 106
Nucleocytoplasmic shuttling, 170, 171
Nucleoprotein receptor (NR), 121

O

2-O-methyl transferase activity, 85
Oncolytic viruses, 147
Orderpromoting residues, 112

P

Paramyxoviridae, 135, 137, 154
Paramyxovirinae, 150, 152
Particle assembly, 64–65
PCT, 107
Peebles, T., 4, 5
Pharmaceutical firms, 7

Phase, 89, 91, 92
Pirh2, 178
Ploidy, 78
PNT, 107
Polarised epithelial cells, 165, 173, 174
Poliomyelitis virus, 131
Polyadenylation, 88, 90
Polyploid, 135
Polysomes, 87, 96
P protein, 79, 81, 84–86, 93, 170, 176, 178, 179
Preimmunity, 151
Premolten globule, 112
Prime-boost, 151
Promoter, 81, 86, 90, 93
Protein self-cleavage, 141

Q
Qbeta, 131

R
Rab9, 172
RdRp. *See* RNA-dependent RNA polymerase (RdRp)
Real-space helical reconstruction, 110
Receptor attachment, 62, 63, 65, 70
Regulatory agencies, 149, 153
Replication, 93–94, 132, 166, 168–171, 174, 177, 180, 181, 183
Replication competent vaccines, 150, 151
Replication deficient vaccines, 150
Replication promoter, 116, 117
Rescue, 130–138, 140, 147, 148, 154
Reverse genetics, 129–155
Rhabdoviridae, 110–111, 134, 135, 137, 154
Ribonucleoprotein (RNP), 78, 79, 81, 84–89, 91, 93, 97, 132–134, 136, 138, 149
RIG-I, 93, 173–176, 183
Rinderpest virus (RPV), 21, 22
Risks, 153
RNA-dependent RNA polymerase (RdRp), 79, 84–86, 88–93
RNaseL, 83
RNA synthesis, 79, 85, 87, 90–92
RNP, 104
RPV. *See* Rinderpest virus (RPV)
Rule of six, 104, 139, 140, 149, 150

S
SARS corona virus, 146–147
Schwarz, J.A., 8
Schwarz/Moraten MV strain, 144
Segmented MVs, 147–148

Signaling lymphocyte activation molecule (SLAM), 15–25
SIV/HIV, 145
SLAM, 60–62, 65, 67–69, 137, 140, 147, 165. *See* Signaling lymphocyte activation molecule (SLAM)
SSPE. *See* Subacute sclerosing panencephalitis
STAT1, 176, 177, 183
Subacute sclerosing panencephalitis (SSPE), 8, 83, 88, 92, 96, 139, 140, 151
Susceptible children, 6
Syncytia, 135
Syncytium, 175

T
TNFα, 177
Toll-like receptors (TLR), 83, 178
Trailer, 84, 93
Transcription, 79, 83–93, 132, 168, 169, 173, 174, 179–181, 183
Transcription gradient, 90–92
Transcription units, 81, 84, 86, 87, 94
Transgenic CD150 mice, 168
Translation, 94–96
Translational control, 94, 95
Trans-synaptic transfer, 96
2,2,2-Trifluoroethanol (TFE), 114
5'-Triphosphate RNA, 173, 174
T7 RNA polymerase, 134–138, 148
Tubulin, 81, 85
Two-dose schedule, 9

U
5' UTR, 95, 96

V
Vaccine, 166–169, 172, 176, 178, 181
Vaccinia-T7 helper virus, 134, 148
Vimentin, 81
Viral factories, 79, 81, 96
Virions, 134, 135
Virus assembly, 171, 173, 183
Virus budding, 172
V protein, 83–85, 88, 94, 95, 176–178, 183
VTF7-3, 134

W
Warthin-Finkeldey cells, 176

Z
Zinc-binding finger, 85
Zinsser, Hans, 4

Current Topics in Microbiology and Immunology

Volumes published since 2002

Vol. 271: **Koehler, Theresa M. (Ed.):** Anthrax. 2002. 14 figs. X, 169 pp. ISBN 3-540-43497-6

Vol. 272: **Doerfler, Walter; Böhm, Petra (Eds.):** Adenoviruses: Model and Vectors in Virus-Host Interactions. Virion and Structure, Viral Replication, Host Cell Interactions. 2003. 63 figs., approx. 280 pp. ISBN 3-540-00154-9

Vol. 273: **Doerfler, Walter; Böhm, Petra (Eds.):** Adenoviruses: Model and Vectors in VirusHost Interactions. Immune System, Oncogenesis, Gene Therapy. 2004. 35 figs., approx. 280 pp. ISBN 3-540-06851-1

Vol. 274: **Workman, Jerry L. (Ed.):** Protein Complexes that Modify Chromatin. 2003. 38 figs., XII, 296 pp. ISBN 3-540-44208-1

Vol. 275: **Fan, Hung (Ed.):** Jaagsiekte Sheep Retrovirus and Lung Cancer. 2003. 63 figs., XII, 252 pp. ISBN 3-540-44096-3

Vol. 276: **Steinkasserer, Alexander (Ed.):** Dendritic Cells and Virus Infection. 2003. 24 figs., X, 296 pp. ISBN 3-540-44290-1

Vol. 277: **Rethwilm, Axel (Ed.):** Foamy Viruses. 2003. 40 figs., X, 214 pp. ISBN 3-540-44388-6

Vol. 278: **Salomon, Daniel R.; Wilson, Carolyn (Eds.):** Xenotransplantation. 2003. 22 figs., IX, 254 pp. ISBN 3-540-00210-3

Vol. 279: **Thomas, George; Sabatini, David; Hall, Michael N. (Eds.):** TOR. 2004. 49 figs., X, 364 pp. ISBN 3-540-00534X

Vol. 280: **Heber-Katz, Ellen (Ed.):** Regeneration: Stem Cells and Beyond. 2004. 42 figs., XII, 194 pp. ISBN 3-540-02238-4

Vol. 281: **Young, John A. T. (Ed.):** Cellular Factors Involved in Early Steps of Retroviral Replication. 2003. 21 figs., IX, 240 pp. ISBN 3-540-00844-6

Vol. 282: **Stenmark, Harald (Ed.):** Phosphoinositides in Subcellular Targeting and Enzyme Activation. 2003. 20 figs., X, 210 pp. ISBN 3-540-00950-7

Vol. 283: **Kawaoka, Yoshihiro (Ed.):** Biology of Negative Strand RNA Viruses: The Power of Reverse Genetics. 2004. 24 figs., IX, 350 pp. ISBN 3-540-40661-1

Vol. 284: **Harris, David (Ed.):** Mad Cow Disease and Related Spongiform Encephalopathies. 2004. 34 figs., IX, 219 pp. ISBN 3-540-20107-6

Vol. 285: **Marsh, Mark (Ed.):** Membrane Trafficking in Viral Replication. 2004. 19 figs., IX, 259 pp. ISBN 3-540-21430-5

Vol. 286: **Madshus, Inger H. (Ed.):** Signalling from Internalized Growth Factor Receptors. 2004. 19 figs., IX, 187 pp. ISBN 3-540-21038-5

Vol. 287: **Enjuanes, Luis (Ed.):** Coronavirus Replication and Reverse Genetics. 2005. 49 figs., XI, 257 pp. ISBN 3-540- 21494-1

Vol. 288: **Mahy, Brain W. J. (Ed.):** Foot-and-Mouth-Disease Virus. 2005. 16 figs., IX, 178 pp. ISBN 3-540-22419X

Vol. 289: **Griffin, Diane E. (Ed.):** Role of Apoptosis in Infection. 2005. 40 figs., IX, 294 pp. ISBN 3-540-23006-8

Vol. 290: **Singh, Harinder; Grosschedl, Rudolf (Eds.):** Molecular Analysis of B Lymphocyte Development and Activation. 2005. 28 figs., XI, 255 pp. ISBN 3-540-23090-4

Vol. 291: **Boquet, Patrice; Lemichez Emmanuel (Eds.):** Bacterial Virulence Factors and Rho GTPases. 2005. 28 figs., IX, 196 pp. ISBN 3-540-23865-4

Vol. 292: **Fu, Zhen F. (Ed.):** The World of Rhabdoviruses. 2005. 27 figs., X, 210 pp. ISBN 3-540-24011-X

Vol. 293: **Kyewski, Bruno; Suri-Payer, Elisabeth (Eds.):** CD4+CD25+ Regulatory T Cells: Origin, Function and Therapeutic Potential. 2005. 22 figs., XII, 332 pp. ISBN 3-540-24444-1

Vol. 294: **Caligaris-Cappio, Federico, Dalla Favera, Ricardo (Eds.):** Chronic Lymphocytic Leukemia. 2005. 25 figs., VIII, 187 pp. ISBN 3-540-25279-7

Vol. 295: **Sullivan, David J.; Krishna Sanjeew (Eds.)**: Malaria: Drugs, Disease and Post-genomic Biology. 2005. 40 figs., XI, 446 pp. ISBN 3-540-25363-7

Vol. 296: **Oldstone, Michael B. A. (Ed.)**: Molecular Mimicry: Infection Induced Autoimmune Disease. 2005. 28 figs., VIII, 167 pp. ISBN 3-540-25597-4

Vol. 297: **Langhorne, Jean (Ed.)**: Immunology and Immunopathogenesis of Malaria. 2005. 8 figs., XII, 236 pp. ISBN 3-540-25718-7

Vol. 298: **Vivier, Eric; Colonna, Marco (Eds.)**: Immunobiology of Natural Killer Cell Receptors. 2005. 27 figs., VIII, 286 pp. ISBN 3-540-26083-8

Vol. 299: **Domingo, Esteban (Ed.)**: Quasispecies: Concept and Implications. 2006. 44 figs., XII, 401 pp. ISBN 3-540-26395-0

Vol. 300: **Wiertz, Emmanuel J.H.J.; Kikkert, Marjolein (Eds.)**: Dislocation and Degradation of Proteins from the Endoplasmic Reticulum. 2006. 19 figs., VIII, 168 pp. ISBN 3-540-28006-5

Vol. 301: **Doerfler, Walter; Böhm, Petra (Eds.)**: DNA Methylation: Basic Mechanisms. 2006. 24 figs., VIII, 324 pp. ISBN 3-540-29114-8

Vol. 302: **Robert N. Eisenman (Ed.)**: The Myc/Max/Mad Transcription Factor Network. 2006. 28 figs., XII, 278 pp. ISBN 3-540-23968-5

Vol. 303: **Thomas E. Lane (Ed.)**: Chemokines and Viral Infection. 2006. 14 figs. XII, 154 pp. ISBN 3-540-29207-1

Vol. 304: **Stanley A. Plotkin (Ed.)**: Mass Vaccination: Global Aspects – Progress and Obstacles. 2006. 40 figs. X, 270 pp. ISBN 3-540-29382-5

Vol. 305: **Radbruch, Andreas; Lipsky, Peter E. (Eds.)**: Current Concepts in Autoimmunity. 2006. 29 figs. IIX, 276 pp. ISBN 3-540-29713-8

Vol. 306: **William M. Shafer (Ed.)**: Antimicrobial Peptides and Human Disease. 2006. 12 figs. XII, 262 pp. ISBN 3-540-29915-7

Vol. 307: **John L. Casey (Ed.)**: Hepatitis Delta Virus. 2006. 22 figs. XII, 228 pp. ISBN 3-540-29801-0

Vol. 308: **Honjo, Tasuku; Melchers, Fritz (Eds.)**: Gut-Associated Lymphoid Tissues. 2006. 24 figs. XII, 204 pp. ISBN 3-540-30656-0

Vol. 309: **Polly Roy (Ed.)**: Reoviruses: Entry, Assembly and Morphogenesis. 2006. 43 figs. XX, 261 pp. ISBN 3-540-30772-9

Vol. 310: **Doerfler, Walter; Böhm, Petra (Eds.)**: DNA Methylation: Development, Genetic Disease and Cancer. 2006. 25 figs. X, 284 pp. ISBN 3-540-31180-7

Vol. 311: **Pulendran, Bali; Ahmed, Rafi (Eds.)**: From Innate Immunity to Immunological Memory. 2006. 13 figs. X, 177 pp. ISBN 3-540-32635-9

Vol. 312: **Boshoff, Chris; Weiss, Robin A. (Eds.)**: Kaposi Sarcoma Herpesvirus: New Perspectives. 2006. 29 figs. XVI, 330 pp. ISBN 3-540-34343-1

Vol. 313: **Pandolfi, Pier P.; Vogt, Peter K. (Eds.)**: Acute Promyelocytic Leukemia. 2007. 16 figs. VIII, 273 pp. ISBN 3-540-34592-2

Vol. 314: **Moody, Branch D. (Ed.)**: T Cell Activation by CD1 and Lipid Antigens, 2007, 25 figs. VIII, 348 pp. ISBN 978-3-540-69510-3

Vol. 315: **Childs, James, E.; Mackenzie, John S.; Richt, Jürgen A. (Eds.)**: Wildlife and Emerging Zoonotic Diseases: The Biology, Circumstances and Consequences of Cross-Species Transmission. 2007. 49 figs. VII, 524 pp. ISBN 978-3-540-70961-9

Vol. 316: **Pitha, Paula M. (Ed.)**: Interferon: The 50th Anniversary. 2007. VII, 391 pp. ISBN 978-3-540-71328-9

Vol. 317: **Dessain, Scott K. (Ed.)**: Human Antibody Therapeutics for Viral Disease. 2007. XI, 202 pp. ISBN 978-3-540-72144-4

Vol. 318: **Rodriguez, Moses (Ed.)**: Advances in Multiple Sclerosis and Experimental Demyelinating Diseases. 2008. XIV, 376 pp. ISBN 978-3-540-73679-9

Vol. 319: **Manser, Tim (Ed.)**: Specialization and Complementation of Humoral Immune Responses to Infection. 2008. XII, 174 pp. ISBN 978-3-540-73899-2

Vol. 320: **Paddison, Patrick J.; Vogt, Peter K. (Eds.)**: RNA Interference. 2008. VIII, 273 pp. ISBN 978-3-540-75156-4

Vol. 321: **Beutler, Bruce (Ed.):** Immunology, Phenotype First: How Mutations Have Established New Principles and Pathways in Immunology. 2008. XIV, 221 pp. ISBN 978-3-540-75202-8

Vol. 322: **Romeo, Tony (Ed.):** Bacterial Biofilms. 2008. XII, 299. ISBN 978-3-540-75417-6

Vol. 323: **Tracy, Steven; Oberste, M. Steven; Drescher, Kristen M. (Eds.):** Group B Coxsackieviruses. 2008. ISBN 978-3-540-75545-6

Vol. 324: **Nomura, Tatsuji; Watanabe, Takeshi; Habu, Sonoko (Eds.):** Humanized Mice. 2008. ISBN 978-3-540-75646-0

Vol. 325: **Shenk, Thomas E.; Stinski, Mark F.; (Eds.):** Human Cytomegalovirus. 2008. ISBN 978-3-540-77348-1

Vol. 326: **Reddy, Anireddy S.N; Golovkin, Maxim (Eds.):** Nuclear pre-mRNA processing in plants. 2008. ISBN 978-3-540-76775-6

Vol. 327: **Manchester, Marianne; Steinmetz, Nicole F. (Eds.):** Viruses and Nanotechnology. 2008. ISBN 978-3-540-69376-5

Vol. 328: **James L. van Etten (Ed.):** Lesser Known Large dsDNA Viruses 2008. ISBN 978-3-540-68617-0

Vol. 329: **Diane E. Griffin; Michael B.A. Oldstone (Eds.):** Measles 2009. ISBN 978-3-540-70522-2

Printing: Krips bv, Meppel, The Netherlands
Binding: Stürtz, Würzburg, Germany